PREFACE

Without much doubt, Numerical Control is the foremost application of the computer in manufacturing.

In 1979 over 9,000 plants had installed a minimum of one numerically controlled machine. These machines were milling or tapping; boring or drilling, as well as completing a host of other applications.

The first numerically controlled machine was produced by the Massachusetts Institute of Technology in 1952. It was created to repeatedly machine aircraft components.

The story of Numerical Control—and its increasing use—parallels that of the rise of the computer—the ''brain'' which translated the seemingly meaningless holes in paper tape into machine action.

In the 1950's computers were large in size, high in price and lacking in reliability. As those three factors changed and as programming became easier, engineers found more applications for NC and the computer in the manufacturing environment.

Today, the machining center, a term unheard of before NC, is capable of performing several operations in one setup. There is, seemingly, an unlimited number of applications which can be accomplished on a machine. How best to use NC is a challenge to be faced by engineers in plants throughout the world.

The purpose of this volume is to outline some of the uses of Numerical Control today. In reviewing these applications, it is hoped that the reader will gain insights and develop new ideas toward his or her NC application challenges.

With the current state-of-the-art in Computerized Numerical Control (CNC), Direct Numerical Control (DNC), Computer-Aided Design (CAD)/Computer-Aided Manufacturing (CAM) and the current, minimal penetration in the market, papers and articles have been selected on the basis of general exposure as opposed to pointed details which may soon be out of date.

Chapter One, NC Moving Forward, presents an overview of NC application and cost, and explains what lies ahead for NC.

Chapter Two, Practical Cases, discusses the applications of Numerical Control in pressworking, shearing, flame cutting as well as in machining large castings and others. This chapter also presents a discussion of programming.

Chapter Three reviews DNC and Flexible Manufacturing. The chapter presents some factors to be considered when evaluating the purchase and use of numerically controlled machine tools. The chapter also reviews some of the outstanding examples of DNC at work and discusses automated batch manufacturing systems—a system for the 1980's.

This volume's final Chapter, Towards CAD/CAM, discusses how the computer is being integrated into the manufacturing process. Articles discuss the use of computer graphics in numerically controlled machining, as well as effective CAM, DNC, CNC and NC System Maintenance.

I wish to thank the authors of each of these articles for their contributions. (Author's titles are those that they held when they wrote the journal article or paper.) I also wish to express my gratitude to the publications who supplied some of the material in this volume. These include: *American Machinist*, the *International Journal of Production Research*, *Iron Age* and *Iron Age Metalworking International*, *Machinery and Production Engineering*, *Manufacturing Engineering*, *Mechanical Engineering*, *Modern Machine Shop*, *Product Engineering*, *Production*, *Production and Inventory Management*, *Production Engineering*, and *Tooling & Production*. My thanks also to: the Publishing Division of Her Majesty's Stationery Office in England, the National Engineering Laboratory in Scotland, and the North Holland Publishing Company in The Netherlands. I also wish to thank Bob King and Judy Stranahan of the SME Marketing Services Department for their help in assembling this volume.

I would like to dedicate this volume to a great number of people. My personal thanks is extended to all of those who had to make their first NC machine "work." They are users, builders, sellers, managers, programmers, machinists, operators, maintenance technicians, tool setters and the hundreds (possibly thousands) who carefully punched the tapes for the programmers. NC is an outstanding success due to their great efforts.

Jack Moorhead
Editor

CASA

The Computer and Automated Systems Association of the Society of Manufacturing Engineers. . .CASA of SME. . .is an educational and scientific association for computer and automation systems professionals. CASA was founded in 1975 to provide for the comprehensive and integrated coverage of the fields of computers and automation in the advancement of manufacturing. CASA is the organizational "home" for engineers and managers concerned with computer and automated systems.

The Association is applications-oriented, and covers all phases of research, design, installation, operation and maintenance of the total manufacturing system within the plant facility. CASA activities are designed to do the following:

- Provide professionals with a single vehicle to bring together the many aspects of manufacturing, utilizing computer systems automation.
- Provide a liaison among industry, government and education to identify areas where technology development is needed.
- Encourage the development of the totally integrated manufacturing plant.

The application of computer/automated systems must always be timely and must assure cost-effective manufacturing and quality products.

CASA is an official association of the Society of Manufacturing Engineers. In joining CASA of SME, you become a partner with over 50,000 other manufacturing-oriented individuals in 35 countries around the world.

A member of CASA benefits from a constant output of data and services, including a discount on all CASA activities and SME books. CASA's educational programs, chapter membership meetings, publications, conferences and expositions have proven valuable by updating a member's knowledge and skills and by expanding his technical outlook in the integration of manufacturing systems.

Membership in CASA is a means for continuing education. . .a forum for technical dialogue. . .a direct channel for new ideas and concepts. . .an important extension of your professional stature.

TABLE OF CONTENTS

CHAPTERS

1 NC MOVING FORWARD

2 PRACTICAL CASES

3 DNC AND FLEXIBLE MANUFACTURING

4 TOWARDS CAD/CAM

CHAPTER 1

NC MOVING FORWARD

The Future of Numerical Controls

M. M. BARASH[1]
Purdue University, West Lafayette, Ind.

Here is a brief survey of the status of numerical control hardware and software.   The probable level of NC technology by the year 2000 is predicted. Millions of machine tools will soon have to be replaced.   Inspection, parts handling, and some management phases will be incorporated into future computer-controlled automated manufacturing systems.

[1] Professor, Industrial Engineering.   Mem. ASME.

Reprinted from Mechanical Engineering, September, 1979

Robots with artificial sensory perception will handle parts in many applications. Here, a Cincinnati Milacron T³ industrial robot is being used in a welding operation under the control of an Acramatic CNC minicomputer-based "brain." The robot is programmed by a simple, hand-held "teach" pendant.

In 1956 the first commercial numerically controlled machine tool became available. That was over 20 years ago. Twenty years from now we will be approaching the year 2000, a date that is the focus of quite a few recent predictions. Let us then speculate on the status of numerical control by the year 2000.

There is little doubt that in two decades millions of machine tools will be replaced, especially in American industry, which still has a high proportion of outdated equipment.

A few Delphi[2] studies into the future of manufacturing were published in recent years, and since they build on one another, the latest [1][3] should be the most interesting. It does not, however, go as far as the year 2000, and rather surprisingly does not attempt to predict the future proportion of NC equipment. Nonetheless, a reader interested in possible developments in manufacturing is well advised to consult that report.

The question of "saturation" with NC equipment has been discussed in the literature, especially foreign. Estimates of the percentage of total machinery installed that NC equipment—primarily machine tools—will then constitute ranges from 15 percent for West Germany to 25 percent for the Soviet Union. Because of its higher productivity, NC equipment, while being 15 percent of the total, is likely to account for 50 percent of all non-mass-produced items. The author believes that by the year 2000 at least 75 percent of all mechanical parts that are not mass produced will be made on NC equipment. This would require a stock of about 500,000 NC machine tools, or an eightfold increase from 1978.

A yearly growth of just over 8 percent in the construction of NC equipment can accomplish this. The corresponding production figure would be about 38,000 NC machine tools a year by 2000, in an economy twice the size of our present one. It need not pose a problem, especially as such a rate of production would be not more than is necessary for replacement, bearing in mind the higher intensity at which NC equipment operates.

Promoting Forces

Technical and even economic feasibility are not sufficient, though necessary, reasons for a development to occur. There must be strong stimuli, which either make the development highly profitable or essential for the conduct of business. So much has been written on the advantages of numerical control that a repetition would be boring. There are, however, two factors, not always adequately stressed, that not only force the spread of numerical control but also necessitate its continuing improvement and refinement. One is the demand for

ever higher product quality; the other, shrinkage of the skilled labor pool. Neither trend is likely to be reversed in the foreseeable future.

Present Status

Numerical, or digital control has long ceased to be confined to metal cutting machine tools. One can hardly find industrial equipment of any kind today that has no digitally controlled version. However, metal cutting, and grinding, machine tools not only dominate the field of digital control in numbers but also in sophistication.

A numerically controlled machine tool can be considered to have three categories of subsystems: passive, actuators, and controls.

The Passive Subsystem. The passive subsystem is the machine proper—its structure, shafts, spindle(s), bearings, gears, tool magazine(s), tool changer(s), workpiece changers, and all other motion transmitting elements. An enormous variety of design configurations of the machine proper exists already and new ones are being created almost daily. There is every reason to expect that this activity will continue indefinitely, especially as integrated systems of NC machines, the so-called computerized manufacturing systems (CMS), become more popular. Digital control has made possible mechanical design configurations that did not exist before. The oldest and best known is the machine with a magazine of tools which are automatically changed on command, the machining center. Today there are turning centers, milling centers, multi-operation (milling-drilling-boring) centers, with one or more tool spindles and a corresponding number of tool magazines and changers. A number of recent multipurpose machining centers change not individual tools but entire heads, each with several tools. The designers' imagination is the only limit to this creative activity.

The quality of mechanical elements of NC machine tools has been steadily improving, and has reached a level never anticipated for conventional equipment. Computer-aided design is extensively employed for optimization of structures, spindles, and other critical elements. Spindles run in superior rolling element, hydrodynamic, hydrostatic, or aerostatic (gas film) bearings. Journal surfaces are often hard coated, sometimes with flame-sprayed ceramics. Hydrostatic and aerostatic guideways are well developed, not to mention rolling element type. Their limited use is due to high cost rather than to technical problems.

To reduce cost while increasing design versatility, leading machine tool builders employ modular design. Welded structures are replacing castings, which greatly reduces lead time in construction. Very ingenious solutions were found for such problems as speeding up automatic tool changing (which in modern machining centers may take as little as 5 seconds and even less),

[2] Delphi Forecast of Manufacturing Technology, Society of Manufacturing Engineers, University of Michigan, 1977.
[3] Numbers in brackets designate References at end of article.

avoidance of interference when indexing multiple tool holders and turrets, better utilization of working space so as to engage more tools simultaneously, etc. Even if further development were halted for two decades, which, of course, will not happen, the latest NC machine tool models would still be quite effective in the year 2000.

Actuators. In the field of actuators the most significant development is emergence of electric motor drives for controlled axes as the dominant, almost exclusive, mode. There is also a growing proponderance of d.-c. motors for spindle drives, except for ultra-high-speed internal grinding and certain special-purpose spindles. Pneumatic and hydraulic actuators, however, have not disappeared, and are employed for clamping, indexing, and other functions where their small size is advantageous.

The expanded use of electric drives has resulted in greatly simplified and cheaper mechanisms, both in control loops and in main drives (such as headstocks). It has also increased machine reliability and simplified its maintenance.

Controls. Even more dramatic progress is observed in NC controls. The old tape reader-cum-special controller has practically disappeared. All but the simplest new NC machines are equipped with computer-numerical-control (CNC), basically a minicomputer which, in most cases, is microprocessor based. The cost of modern CNC is quite low compared to controls of even a decade ago. Moreover, it offers extensive additional capabilities which are only beginning to be realized. One of the latest CNC systems employs a floppy disk which is loaded from a large computer. Tape is eliminated.

In addition to executing the NC program, CNC systems are being used to keep track of machine activities, to operate an adaptive control loop and to monitor some of the machine "vital signs" such as spindle bearing temperature. In most recent CMS and DNC (direct numerical control, i.e., control of several machines by one large computer) the CNC system serves as interface with the main computer.

CNC offers the opportunity of exercising control over the machining process. Adaptive control was mentioned earlier. Commercially available systems measure forces and deflections in milling, drilling, and turning operations, and modify process variables like speed and feed to minimize, within constraints, a certain objective function such as production time or cost. The latest CNC-based adaptive control system for cylindrical grinding minimizes the time or cost required to produce a part to specified diameter and surface finish. These are only a few examples of the power of CNC. Many new uses will come with time.

Measurement and Inspection. Laser sensors for direct noncontact measurement of dimensions of parts being machined are under development. Laboratory prototypes are already working. A go/no-go electronic bore gage that sends its signals by radio has been developed and will be on the market shortly. Such a gage can be handled as just one of the tools in the magazine.

If the cutting tool is replaced with a probe, and suitable design changes are made, an NC machine tool becomes a digital measuring and inspection machine. Its output is a listing of dimensions checked according to an NC program. Thousands of digital inspection machines are in use worldwide. Some are driven by tape, the latest by a minicomputer.

Materials Handling. A major factor in productivity is the loading and unloading of parts to and from the machine tool. In some cases robots are employed for this purpose. They range from simple mechanical arms to full-fledged multiple axis units. In either case, they can be easily controlled by the CNC system of the machine tool.

Computerized Manufacturing Systems. The latest development in the field of numerical control is the computerized manufacturing system (CMS), also called flexible manufacturing system (FMS). Such a system consists of a group of NC machine tools, other equipment such as a digital inspection machine, a washing unit, etc., and an automatic materials handling system that links all of the mentioned operating stations.

Control of the entire facility is by a computer, which contains all NC part programs as well as routing programs. These systems have the versatility of general-purpose machine tools combined with efficiency close to that of a dedicated transfer line. There are some 30 systems worldwide and a half a dozen in the U.S. employed in batch production. A typical system with six NC machines may have six to 10 different parts in process at the same time, and the batch size for each part may be anywhere from one to several hundred. CMS are quite expensive, sometimes costing several million dollars in equipment alone, and their effective operation is of prime importance. A system that is well operated can outproduce stand-alone NC machines by 50 percent and more, and do so with fewer people. Optimal design and operation of such systems is being studied under NSF grants at Purdue University and MIT.

Software. One cannot review the status of NC without mentioning the relevant software. To produce a shape the cutting tool must be guided along a path which is related to the desired shape. Commands for controlling tool motion are generated through transformation of the mathematical description of the shape in question. The description is entered, using appropriate language, into a computer program which generates the sequence of tool motion controlling signals. There are nearly 100 computer languages worldwide for describing part geometry and producing the necessary NC machine in-

Computerized manufacturing system developed by White Sunstrand Machine Tool, Inc., Belvidere, Ill., incorporates floor mounted rails and a pallet carrying shuttle cars for material handling. As seen in the system arrangement diagram, there are: three Sunstrand four-axis OD-3 drilling machines. A; two vertical turret lathes, B; four Sunstrand five-axis OM-3 Omnimil machining centers, C; and a DEA coordinate-axis inspection machine. The photograph shows a White-Sunstrand Omniline manufacturing system with the work load-unloaded area in the foreground. The system is entirely computer controlled.

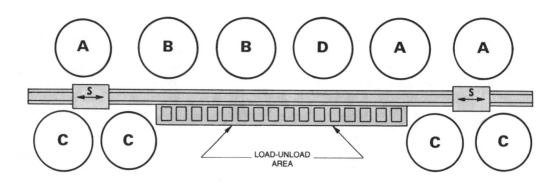

LOAD-UNLOAD AREA

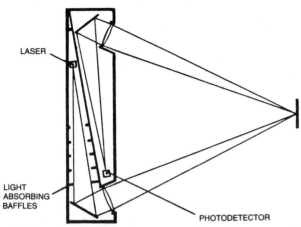

LASER

LIGHT
ABSORBING
BAFFLES

PHOTODETECTOR

In-process optical gaging for numerical machine tool control and automated processes is under development and laboratory proto-types are already working. Here's a possible configuration of an optical gaging head using a solid-state laser, as reported by H.W. Mergler and Steven Sahajdah of Case Western Reserve University.

Special machine tools employing the latest concepts in numerical control are featured in this Headchanger system built for the I.J. Case Co. by the Ingersoll Milling Machine Co. of Rockford, Ill. From the console, left, the operator controls the load/unload conveyor for pallets, the indexing units, and the change head transfer system. The three heads on the left are for drilling and the two on the right are for tapping. Palletized castings are moved on the shuffle conveyor to the four-position index unit under control from an auxiliary panel.

Flexible manufacturing system built by Kearney & Trecker Corp. is capable of randomly process-ing eight different parts at an aggregate volume of about 20,000 parts per year. It features 10 machine tools including five Milwaukee-Matic Modu-Line machining centers, one three-axis N/C rough milling machin-ing center, and four duplex head indexers, each of which stores about 20 randomly selectable heads. Up to 23 Tow-Line carts travel randomly along the path-ways. Loading, unloading, and partial inspection is done at three work stations. Arrows indi-cate direction of cart travel. The system again is entirely compu-ter controlled.

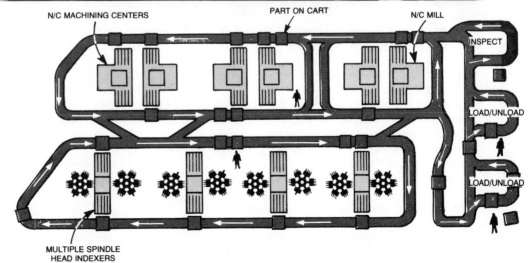

N/C MACHINING CENTERS · PART ON CART · N/C MILL · INSPECT · LOAD/UNLOAD · LOAD/UNLOAD · MULTIPLE SPINDLE HEAD INDEXERS

structions. The most powerful of these software systems is APT—Automatically Programmed Tools, developed in the U.S. during 1956–1970. APT uses terminology of analytic geometry; shapes are described by equations. This includes lines, conic sections, planes, cylinders, cones, spheres, and quadric surfaces. Tool motion commands are produced for up to five axes simultaneously (three translation and two rotation). The system requires a fairly large computer and is needed in only about 5 percent of cases of NC form milling.

Most of NC programs are written not in APT but in simpler languages such as COMPACT-II and others which are easier to use.

APT and most other general languages do not provide process technology, such as the appropriate sequence of cuts, size of cut, cutting speed, method of fixturing, etc. These are entered by the programmer. More recent developments are languages that automatically determine process technology parameters. A number of such languages have been developed for turning; most are tied to specific machine tool makes, but one or two are general. Work in developing these "process planning" language is being pursued in most industrial countries.

Very recently there have appeared, both in the U.S. and abroad, CNC systems (so far only for turning machines) that allow programming directly at the console, using a simple language. The language is so designed that anyone who can read an engineering drawing is able to use it (unlike the analytical geometry-based languages of the APT class). In some cases the CNC system even generates the process plan, allocates cuts distributions, and specifies feeds and speeds. There is little doubt that this and similar developments will play a major role in the introduction of NC to medium and small machine shops. Still, it is probably fair to say that status of NC software, in the broad sense, is years behind the hardware.

The Future

In speculating on the future of numerical control one has to agree to a number of assumptions:

1 Metal removal will maintain its position among processes, though the proportion of grinding may go up, and of machining, down.

2 The structure of the metalworking industry will remain largely unchanged, i.e., there will still be a large number of medium and small enterprises.

3 Computer power per dollar will be two orders of magnitude greater than today.

4 Time sharing networks will be more numerous and more powerful.

It is to be expected that there will be an extensive spread in the age distribution of the NC equipment population. Some of the machines built today will still be working, but more than half will be less than 10 years old.

It is to be expected that improved cutting tools and abrasives will be developed in the next decade, and, therefore, by the year 2000 machine tools will operate at speeds on the average twice as high as today. Appropriate designs of spindle bearings, chucks, bar feeders, and other speed-affected elements will perforce be developed. New tooling schemes will be used, especially in high-volume machining to combine several operations and to minimize the effect of cutting forces on product accuracy.

The average accuracy required of machined parts will be higher than today, and many machine tools will be equipped with adaptive control systems which will monitor forces and temperature in various parts of the machine as well as workpiece dimension and will compensate for both slowly and fast changing deviations. There will be on-line tool wear sensors which will enable the adaptive control system to optimize the machining process. In addition, critical elements of the machine will be monitored so that impending failure will be detected well in advance. The CNC system of the machine will be of such power that it will easily cope with all the required computing tasks.

Many machine tools will fully inspect the parts they make, using telemetry type gauges, automatic laser interferometers, and other advanced metrology tools.

Manipulators and robots will be extensively used to load and unload parts and to transfer them from machine to machine. Except for very heavy and/or complex parts, palletizing or fixturing will be fully or partly automatic. Numerous types of automatic, digitally controlled fixtures will be in use.

The majority of all NC machines will be integrated into manufacturing "cells," with automatic materials handling where feasible. The cell will be a production unit for which a common plan will be made resident in a supervisory computer. This computer will instruct the individual machine CNC systems and "listen" to their reports. In turn, the supervisory computer will be reporting to a management computer.

About a tenth of all NC machines will be included in fully integrated CMS with completely automatic operation, including materials handling. There will be about 5000 such systems by the year 2000. At least 100 of these will be further integrated into about 20 super-systems which will include automatic, programmable assembly of parts made by the "member" CMS into near-finished products. All CMS arrangements will be equipped with instrumentation for tool wear and fracture sensing, adaptive control, inspection, machine condition monitoring, automatic tool delivery subsystems (in addition to local magazines), and in-process automatic storage. Prior to reaching the automatic assembly section parts will be automatically "unfixtured," washed, dried, marked, chemically treated, and reinspected. Fixtures and pallets will be delivered by a separate automatic handling system to load stations which will be in most cases automatic. Manipulators and robots with artificial sensory perception (optical, acoustic, microwave, electromagnetic, tactile-mechanical) will handle parts in load-unload and assembly stations.

Software for NC manufacturing will be greatly developed above the present level. There will be a few very large systems, easily accessible on a time sharing basis which will be able to automatically design about 75 percent of all mechanical parts. These systems will prepare for them a menu of elective optimal manufacturing processes, with full NC part programs which will also include all tooling and fixturing data, post-processed for any given machine. In addition, there will be quite a few smaller, interactive systems for part designing, process planning and NC programming. For complex parts these systems will require some human assistance.

The large number of management computers in shops and factories will make it possible to collect vast amount of information about types of tools, fixtures, machines, work materials, their performance, and interaction. Some organization will actively process these data banks to create a manufacturing learning system, which in turn will be employed to update reference values in working programs. One result will be a standardization of fixtures, which in turn will simplify automatic process

planning and creation of NC programs without danger of tool-fixture interference. And in general, there will be in use sophisticated methods for program verification which will eliminate the need for trial runs on the NC machine. Computer simulation will be widely used as a daily routine, and some manufacturing systems—especially CMS—will be represented, for simulation purposes, in the form of chips as add-on to the system control computer. Simulations of operating variants will be produced during system normal work.

The above prognostication does not detail new uses of NC or developments in nonmachining areas. These will, of course, happen as well, Digital control will pervade all manufacturing spheres, be it electric machining, forging, sheet metal work, casting, molding of plastics, woodworking, textiles, or printing, just to mention areas first coming to mind.

However, more serious, though inevitable, is the omission of entirely new, now unknown, concepts.

Two decades is long enough for a new idea or discovery to have a significant impact even in such a slowly changing field as machine tools. As to this, one can only express the hope that indeed another great idea will emerge, proving the presented speculations to be conservative.

Based on a paper contributed by the ASME Dynamic Systems and Control Division.

Reprinted by the Society of Manufacturing Engineers from IRON AGE, March 19, 1979; Chilton Company; 1979

NC-CAD/CIM REPORT

A Management Guide to Computer-Integrated Manufacturing, by Alice M. Greene

This month: After the acceptance of numerical control, comes an interest in optimizing its operation on a shop floor. This produces a number of interesting trends.

Charles Britton,
Cincinnati Milacron

"There's a developing awareness of the control as a piece of equipment in its own right, not just as an attachment to a machine tool."

Putting NC to work—more simply

The area of numerical control is crowded with things to talk about and trends to track. This month control builders and machine tool builders shared with us some of the trends they see.

One to which both groups are reacting comes from the users of numerically controlled machine tools. It is the growing tendency of the user industry to standardize on the number of control builders whose equipment they will use within their shops. The main reason attributed to the user viewpoint is simplicity. It simplifies multiple ways of programming different kinds of units, storing multiple sets of parts for different kinds of units, multiple operating procedures, maintenance procedures. It simplifies service to an extent. All these factors can add to the cost of running a machine shop, directly or indirectly. And some users are paying increasing attention to this kind of cost reduction.

Charles Britton, marketing manager, Electronic Systems Div., Cincinnati Milacron Inc., attributes this to "a developing awareness of the control as a piece of equipment in its own right and not just an attachment to a machine tool. It's an indication that the industry is growing up in terms of electronics on the shop floor."

"We are finding more industries who, when buying a certain machinery configuration, are focusing more attention on the controls applied to the machines. Maybe they will select three or four suppliers of control equipment," notes Mr. Britton.

"Three or four suppliers allows a competitive situation to still exist," he feels, "It allows the introduction of features, allows the flexibility needed. But it's better than having 13 different suppliers of equipment which then become disruptive in terms of maintenance and operation." This kind of selection process has been going on primarily in the aerospace industry, and we expect it to continue Mr. Britton adds. You don't find every company in the NC marketplace in this kind of condition. But it is a trend that is developing with the larger users, the ones who have lots of control equipment.

We heard variations of this same story from others. With controls from only a few suppliers, customers tend to know the controls better, they can get service easier, they know where to go for parts, comments T.N. Tellor, vice president, The Warner & Swasey Co.

This situation obviously works to the advantage of the big control builders, both the captive and the general control builder. And each has his own way of looking at the subject.

Mr. Britton states that Cincinnati Milacron, the captive builder with far more shipments than any other, has a very strong position relative to this requirement. "When you talk about supplying electronic equipment to a machine tool facility, not

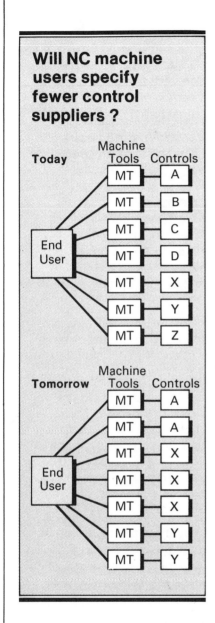

Will NC machine users specify fewer control suppliers ?

only does Cincinnati have the ability to offer standardization of both products and services, but that expands out to a commonality of equipment used in grinder controls, turning equipment, machining centers and aerospace profiling equipment."

So the captives (Cincinnati and others) feel confident of their strength as single source supplier of machinery and controls that, for the most part, have operational and software similarities to each other.

The general controls builders, who acknowledge the trend to control standardization, feel the captive companies will be at a disadvantage because their captive nature precludes the competition necessary to remain innovative. And, therefore, the captives will have to yield more to other controls on their equipment—when an end user wants to standardize on state-of-the-art controls.

Mixing modularity with standardization

There's another interesting twist in this situation. When you couple the trend toward modularization of machine tool controls with the trend toward limiting the number of control builders used (if this becomes more widespread), what's the result?

As Juris Vikmanis, vice president, general manager, Industrial Control Div. for Bendix, sees it the modularity trend may make it more attractive for the captive control builder to start considering an alternative.

"Control builders (non-captive) will be willing to put in their little card rack and the machine tool builder can call it by his own name if he chooses. Bendix builds controls for more than one customer, Ex-Cell-O Corp. is one example, with the customer's name on them," comments Mr. Vikmanis.

"In such a case, we do everything but to the machine tool builder's specifications. Making it customized takes time and talent and we as numerical control builders have that talent." It helps the machine tool builder, adds Mr. Vikmanis. If he has a machine with a great number of options which not everyone will buy, we can provide a very easy way to lock out certain features and lock them back in. Since these features are generally now all in software, this gives the machine tool builder flexibility in marketing.

Mr. Britton views this development as putting the general supplier of controls into a captive-like relationship with the machine tool builder. The physical packaging of the control system itself may present an area of difficulty for any machine manufacturer in picking a generally available supplier of controls.

"Now when the maker of the Brand X machine chooses the Brand A control," as Mr. Britton puts it, "he may eliminate his ability to use Brand B or Brand C controls systems, and may, in fact, develop a very captive-looking relationship with the general supplier of controls. In the long term, everybody may look captive if that trend continues."

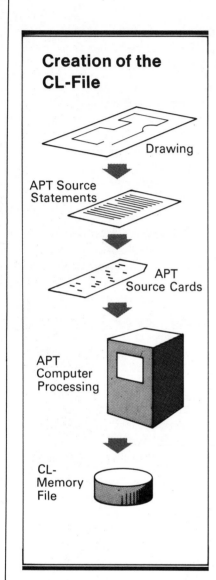

Creation of the CL-File

Drawing

APT Source Statements

APT Source Cards

APT Computer Processing

CL-Memory File

A variation of this relationship is described by T.C. Smith, vice president, General Numeric Corp. "I think long term, control builders will be building hardware and some software and a lot of the software will be controlled by the machine tool builder. The machine tool builder, particularly the large builder, will buy basic hardware, microprocessor-type control with some memory and a basic executive program that says 'this is a lathe' or 'this is a milling machine'.

"Any of the refinements to the executive (that say 'this is a Brand A lathe' or 'a Brand B machining center') will be put together by the machine tool builder. This is being resisted today. But realism says it's coming rapidly."

The end of the postprocessor

Another facet of the move toward simplification is the use of CL-DATA (cutter-location data). This is data which is machine independent. This development, according to L.R. Herndon, Jr., president, Vega Servo-Control, Inc., heralds the end of the postprocessor by transferring the functions of the postprocessor to the control itself.

"In large APT installations today, the postprocessing costs represent typically from 30 to 60 pct of the part programming data processing costs. Not only is the operating cost substantial for each part program postprocessed, but the part program is usable on only one control/machine combination.

"It is no longer necessary to install a new postprocessor on the part program processing computer every time a new machine tool is acquired. The machine dependent postprocessor is not required for the new generation of CL-DATA CNC controls. CL-DATA controls have been installed on a variety of machine tools and are now in production operating on part programs converted by a single CL-converter," says Mr. Herndon.

Cincinnati Milacron has also implemented the transmission of CL-DATA from the APT language or some other geometric language processors straight to the control for some users, according to Mr. Britton.

"This gives the user the ability to redirect the work flow through his shop, if he has equipment that's unavailable at any given time, without having to go through the postprocessing function. It also eliminates the giant storage libraries that result if he has to save programs for every type of machine tool that's on his shop floor in order just to rechannel his work flow.

"The program looks the same, for example, to produce a small part on a small machining center that has fairly low feedrate capabilities or on a large machining center that has very high feedrate capabilities. You would be able to transmit the same program to the control systems of both of those machines, and they would take care of the adapting process to produce the same geometric part and still take advantage of each machine's unique capabilities," states Mr. Britton. □

Reprinted from Manufacturing Engineering, March 1979

1. PAIRED COMBINATION of Workcenter machine (left) and vertical turning machine performs all operations required on torque converters.

Versatile NC Machines Reduce Manufacturing Costs

ROBERT E. GUENGERICH
Plant Manager, Hyster Co.
Sulligent, AL

Increasing production requirements for drive-train components have been met with lower manufacturing costs by using NC machining centers and vertical turning machines

WHEN HYSTER CO. began operations in Alabama in 1970, our first plant at Beaverton was a converted 20,000 ft² (1860 m²) building where we trained a number of employees to machine several components used in the drive-trains of lift trucks. By May, 1971, we had developed the nucleus of a promising workforce, and broken ground for a new, larger plant in nearby Sulligent. With the completion of the new building later that year, operations were transferred to Sulligent and production was started.

Since then, the plant has been ex-panded to a total of 200,000 ft² (18 600 m²), employment has grown to 530, and production has been broadened to include more drive-train components. These components include transmissions, drive axles, steering axles, and hydraulic control valves, which are shipped to various Hyster assembly plants.

Machine Requirements. As production of these parts began, it was recognized that it would be impractical to machine them on conventional equipment and meet production requirements. With production runs varying

2. WORKCENTER MACHINE, used in combination with a vertical turning machine, produces cast aluminum transmission housings.

from 70 to 500 parts, it soon became apparent that the repeatability and flexibility needed would best be served by a number of numerically controlled machines.

Following an intensive analysis of current and future machine tool needs, we began a purchasing program that matched the capabilities of NC and conventional machines with our expanded needs in part production. This continuing program, using capability, price, and delivery as criteria, has been responsible for the purchase of 23

numerically controlled machines, in addition to considerable conventional equipment installed.

Almost one-third of our NC equipment is machining centers from the Workcenter Div. of Ex-Cell-O Corp. These include five 208 machines, a model 108B, and a 408 machine. Two Ex-Cell-O 454 NC vertical turning machines, each with a five-position turret, have also been paired with two of the Workcenter® machines to serve as complete machining centers for specific part families. Because each of these

machines is dedicated to specific part production, tool and fixture changes are kept to a minimum. Also, they are used as production machines simply by changing tapes for differences in part size.

Machining Converters. To machine cast iron torque converters, one of the paired combinations of a Workcenter machine and a vertical turning machine, *Figure* 1, is used. The machining center used in this operation is the only one of the seven without a pallet changer, primarily because machining time

3. MACHINING CENTER mills, drills, and bores both cast steel and iron transmission housings. Critical bores are held to ±0.0005".

4. CAST IRON speed reducer housings and aluminum flywheel housings are both milled and bored on this machining center. ▶

is minimal, and because the fixture permits fast, convenient loading of the casting.

To machine the torque converters, the Workcenter machine is used to mill the location surfaces and drill location holes. Following this operation, the part is fixtured on the vertical turning machine, and the transmission mounting side is rough and finish faced, bored, and drilled. The part is then turned over to face, rough and finish bore, and chamfer the motor mounting side. In producing these parts, a tolerance of ±0.0005" (0.013 mm) is held on the bore size, and a two-position tolerance of 0.002" (0.05 mm) true position (equivalent to 0.007" — 0.018 mm) is maintained on the bores.

Aluminum Housing Production. Across the aisle from these two machines is another paired combination of a Workcenter machine and a vertical turning machine. These are used to machine cast aluminum transmission housings. Production of these parts begins on the machining center, *Figure* 2, where the mounting face is milled and two dowel holes are drilled and bored. The table is indexed 180°, and, after the milling of two faces, two holes are drilled and bored, and the center bore is roughed, finished, chamfered, and back bored.

At the completion of this cycle, the pallet changer moves another casting into position, and the machined part is unloaded and refixtured on the vertical

machine. After locating on the mounting face, the inner surfaces of the part are turned, faced, bored, and chamfered. Bore size of the finished parts is held to ±0.0005" on both machines, and a true position of 0.002" is maintained. The housings go directly to assembly after machining.

Other Machining Operations. A bit further down the aisle, a third Workcenter machine, *Figure* 3, is used to machine cast steel transmission housings for electric lift trucks. After the machine mills the locating surfaces and drills the location holes on the raw casting, it then roughs and finishes all of the housing's critical bores to a tolerance of ±0.0005" and a true position of 0.002". This unit is also used to machine cast iron transmission housings for which bore sizes are held to a tolerance of 0.0003" (0.008 mm). Even with this critical size requirement, production is predictable.

A fourth Workcenter machine, *Figure* 4, is used to produce cast iron speed reducer housings and aluminum flywheel housings. For both parts, we start with qualified castings, and the machining center is used to rough and finish mill the mounting faces, and rough and finish the critical bores of the housings. Again, a tolerance of ±0.0005" on bore size and 0.002" true location are held.

Controls Used. All four of these Workcenter machines are equipped with Ex-Cell-O NCS hard-wired con-

trols with magnetic core memory and buffer storage. They feature three-axis contouring capability along with editing capability for modified tool speed, feed, length, and radius, which can be edited and stored without tape alteration.

Our fifth and most recent Workcenter 208 machine is equipped with a Bendix System 5 controller. This unit, used to machine shifting towers for transmissions, as well as picking up the overload from other machines, offers four-axis contouring capability, alphanumeric readout, block buffer storage, and self-diagnostic and editing capabilities.

Machine Capabilities. All five of the Workcenter 208 machines have a cube size capacity of 24" (610 mm), and can handle workpieces weighing up to 2000 lb (907 kg). Tool changes from the standard 24-tool, random-select, bidirectional conveyors to the 15-hp (11.2-kW) spindles are made in only five seconds.

The lineup of NC machining centers also includes a Workcenter 408 machine, and the most recent addition, a Workcenter 108B machine. The 408 machine, similar in design to the 208 models, handles parts within a 36" (914 mm) cube and weighing up to 5000 lb (2268 kg). It is used primarily to machine cast steel steering axles. The Workcenter 108B machine, which has a highly-sophisticated CNC system mounted directly on the machine column, although not fully operational yet, is being teamed with an NC vertical lathe to machine stub axles.

Advantages. We have found that the use of numerically controlled machines, particularly in our geographic area, has created a number of major benefits. First are the lower manufacturing costs achieved through savings in machining time. Second is the increased versatility. When changing from one part to another, it's a simple matter to change tapes and tooling without incurring excessive downtime. This just isn't possible with conventional equipment.

Third, and last, our plant is located in a rural area of Alabama, where trained journeymen are almost nonexistent. Our employees have impressed us with their ability to learn and their willingness to work, but most of them had no prior machinist training. For that reason, the accuracy and repeatability of NC machining has enabled us to meet our production targets while developing a skilled labor workforce.

As we continue to expand both our plant facilities and the scope of our operations, we feel sure that numerically controlled machines will take on an ever-increasing portion of our parts production. We look to NC to provide many of the solutions to our long and short run production problems, as our experience over the past five years has convinced us that we are only on the leading edge of a new wave of technology that will increase productivity. ∎

Reprinted from Modern Machine Shop, August 1977

Winning With Creative NC

Some numerical control users go strictly by the book. Others discover many operating benfits by applying their own creative ideas.

KEN GETTELMAN, Associate Editor, interviews
INGO WOLFE, Chief Engineer
DON LAWSON, Plant Manager
DICK HARLING, NC Programming Manager
Kurt Manufacturing Company
Minneapolis, Minnesota

Any successful business has its unique presence and mystique that are not always clearly definable, but they are known to its employees, customers and visitors. For example, Kurt Manufacturing Company has become one of the nation's leading contract machine shops with the ability to produce very-tight-tolerance computer components. There are 190 people, 140,000 square feet of floor space, and some 35 NC machines in its active equipment roster.

When talking with three of Kurt's operating managers one afternoon, some of that unique company personality began to emerge. If there is any single word that describes this very successful contract facility, it would have to be "creativity."

Creativity does not mean recklessness. But it does mean taking the basic and proven methodologies and disciplines and branching out from them with new and innovative ideas—ideas that cut production time and costs, and give people a greater sense of accomplishments. It means a certain amount of calculated risk, but it also means a growing list of satisfied customers who get contract machining services faster, with better quality, and at a reasonable cost.

Our discussions centered on four areas of creative innovation. None of them is beyond the ability of any competent NC equipment user—they simply require a willingness to go beyond a narrow textbook approach to manufacturing. Naturally, they would have to be adapted to the facility's own requirements since no two shops are ever exactly alike.

The four areas that have worked so well for Kurt Manufacturing Company include the workcenter

Don Lawson (left), Ingo Wolfe (center) and Dick Harling plan a new workcenter production process for the precision machining of a cast aluminum workpiece.

concept, the use of pick-off jigs, special machine attachments and modifications, and multiple piece setups on NC machine tools.

A few background questions got the discussions underway, but the heart of the operating philosophy soon began to emerge.

How many NC machines do you have on your current roster?

Mr. Lawson: At the present moment, about 35 and they range from two-axis units to four-axis machining centers.

You say 35 at the moment. Does this mean that the number fluctuates from time to time?

Mr. Lawson: Definitely. Although some of the larger units are somewhat permanent, the fact remains

that we do purchase machines and then will sell them as our requirements change. The equipment inventory and plant floor layout are constantly changing according to our workload. There are a few NC machines that have been purchased and sold by us as many as three times.

How do you handle the programming of workpieces for that many different NC machines?

Mr. Harling: We use both manual programming and the APT processor language on a time-sharing basis. In addition to myself, we have three programmers to handle the workload, which varies from runs of ten or more workpieces to production runs of many thousands.

One of the more noticeable features of your plant layout is the absence of grouping machines by function.

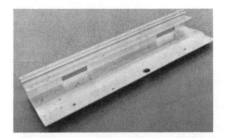

Fig. 1—This rear rail for a computer is machined via the workcenter concept, which means different machine tools, grouped in a common area, are involved in the process.

Fig. 2—The workcenter that is used to machine the workpiece shown in Figure 1 consists of: (clockwise starting at the machine in front of the operator) an in-house-built milling machine to machine the top edge of the rail, a Bridgeport milling machine equipped with side heads to machine the part to length, an NC Cintimatic to mill the two windows, and a modified Natco drilling machine to drill and tap the holes.

Would you explain your philosophy and objectives?

Mr. Wolfe: We are very strong on the workcenter concept, which is designed to create a "flow-through" method of machining a large quantity of parts to avoid stacking and rehandling parts in process.

Could you describe an actual example of this concept?

Mr. Wolfe: An excellent example is the rear rail, shown in Figure 1. It starts out as an aluminum extrusion and goes through a workcenter of four machines and two operators as shown in Figure 2. The first machine was designed and built in-house for this and similar applications. It is a bed-type milling machine with upright columns and side-head mounting capabilities. To mill the top edge of the rail, we mounted a "Precise Super 80" 24,000 rpm milling head over an air-operated fixture. The fixture is tied in with the machining cycle so that the clamping activates the automatic cycle.

The workpiece then goes to a Bridgeport milling machine to which we adapted a T-ram with side heads

for machining the part to length.

From here, the workpiece goes to an NC Cintimatic for milling the two windows. There are two Anglock vises on the Cintimatic table so that we can load one vise while the workpiece in the other vise is being machined. The cutting tool is our own adaptation of an end mill. We ground a point on the end so that the tool will drill a hole in the workpiece and then proceed to mill the window.

After completion of the milling work, the workpiece is moved to a modified Natco drill that is equipped with both drilling and tapping heads and a dial index table. Here, the drilling and tapping of all holes complete the workpiece.

The workpiece seems ideal for complete machining by NC in a single setup. Why do you use a workcenter of four machines?

Mr. Wolfe: If we were doing just a few parts, that would be the preferred method. But with this long run, it definitely pays to use NC only for the window milling and move the

parts through much faster. In other words, we engineer the equipment to the job.

Since this workcenter arrangement was established especially for this job, what happens when the job is finished?

Mr. Wolfe: The machines will either be integrated into another production arrangement, or they will be sold or modified for other work. Machines do not have a fixed arrangement in our plant.

This has strong elements of group technology. Is it necessary to have one operator for each machine to obtain maximum workcenter efficiency?

Mr. Wolfe: No, and this is one of the advantages. One operator handles the first two machines and another operator handles the last two. The work is balanced so that both men make a maximum contribution at a high level of personal involvement, which works out best in terms of efficiency and job satisfaction.

Does the workcenter concept require a completely automatic machine

Fig. 3—A window-frame fixture is used to allow machining of both sides of a workpiece with one setup. Small pick-off jigs mounted on the sides of the workpiece enable the operator to drill and tap less critical holes as the NC machine produces another part.

cycle?

Mr. Lawson: Basically, this is true and this is one reason we now secure NC machines with automatic tool changers. It is not the time saved in changing the tool that makes the automatic tool changer so valuable. It is the free time the operator now has to perform other functions in the work-center arrangement.

You indicated that pick-off jigs were another element of improved machine utilization and operator involvement and satisfaction. Could you cite an example?

Mr. Wolfe: Figure 3 shows a full-contouring machining center with a window-frame fixture, which allows the machine to work on both sides of a workpiece. Since the machining time for this workpiece is quite lengthy, we can more deeply involve the operator in a second operation while saving overall processing time. Four different pick-off jigs can be

seen on the workbench and one is mounted on the workpiece. These simple pick-off jigs, which locate on a previously machined reference point, enable the operator to do some drilling and tapping that could be made part of the NC programming cycle. This methodology involves the operator between workpiece loading and unloading on the machining center and increases the efficiency of the whole setup. It is a very good technique to use when long machining cycles are involved. It also illustrates the value of automatic tool changing to provide free time for the operator to perform other duties.

You also do not hesitate to make extensive machine modifications or make use of attachments on your NC equipment. Do you have a good example of a modification?

Mr. Wolfe: One of the better ones is the modification made to a Cintimatic, as shown in Figure 4, for ma-

chining a large quantity of workpieces. Three special milling spindles were placed on this unit. One spindle carries a small combination side milling and slotting cutter that will mill a 0.040-inch-wide groove along the entire length of the workpiece. The top edge of the groove is also milled with the same cutter. Another spindle is used to mill a pocket with a ⅜-inch-diameter carbide cutter. The third spindle works much the same as a boring bar to create a 30-degree ramp on the part. The allowable tolerance is a total of 0.002 inch on the ramp. By utilizing a 6-inch-diameter cutter, we use only 0.0006 inch of the tolerance.

Each of two fixtures carries both a right- and a left-hand workpiece (they are utilized in pairs); thus, both parts are completed with each NC cycle. As the workpieces in the second fixture are being machined, the operator unloads the workpieces from the first. The methodology has two significant benefits: (1) equal

Fig. 4—A Cintimatic was equipped with three spindles. The vertical spindle on the left is for milling and slotting while the other vertical spindle is used for pocket milling. The horizontal spindle is equipped with a 6-inch-diameter cutter for milling a 30-degree ramp.

amounts of both parts are maintained and (2) the right- and left-hand parts are perfectly matched because both are machined from the same reference with the same tools and a mirror image NC program.

What other benefits are obtained with the special spindles?

Mr. Wolfe: Many of our workpieces are made of aluminum; thus, machining rates can be quite high. By having 24,000 rpm, we can employ fast feed rates and still keep the chip load per tooth at a reasonable level. In fact, feed rates are six times those that would be possible with a standard spindle. By having the three separate spindles mounted close together, the travel time from one to the other is minimized.

You have also made good use of special attachments. Do you have a favorite example?

Mr. Wolfe: Figure 5 shows what can be done with an NC machine when one is willing to adopt special tooling ideas. In this instance, a right-angle milling attachment is being belt-driven from the machine spindle. With this attachment, some very close slots were generated with some critical spacing tolerances. The workpiece was mounted on a vertical index table to present the various sides to a carbide cutter.

One of your other techniques for optimizing NC operations is the use of multiple setups. Why do you favor them?

Mr. Wolfe: The basis for this technique is shown very clearly in Figure 6. Six workpieces are mounted in a multiple-station fixture on a Brown & Sharpe vertical-spindle machining center with automatic tool changer. It was a case of determining the total number of workpieces versus the table-movement time and the tool-changing time. Since the workpieces were small and numerous tools were used on each one, the most economical approach was to move any given tool across all six workpieces before tool changing. Notice that quick-action clamps are used. The operator can start changing workpieces when the last tool is working on the last part. Consequently, shut-down for part loading is minimal when weighed against the number of workpieces being produced.

Being a contract shop, do you find that a repeat order might not be produced on the same NC machine as the initial production run? If so, do you have to completely reprogram the workpiece for another machine?

Mr. Harling: No, we keep the input data for the workpiece, which applies to any machine. Then we only have to go back to the computer for post processing—not a whole new

processing

What are your views on speeds and feeds? Do you follow a conservative path?

Mr. Harling: No, we do not. Actually, we program on the high side and rely on the operator to use the feed rate override if the speeds and feeds are too fast for the material.

Mr. Lawson: Here you have touched upon a very significant point. There are companies that keep operator involvement to a minimum. As was indicated earlier in our conversation, we like maximum operator involvement. It works much better. In fact, it has been our experience that successful NC operation involves the interaction and cooperation of all associates—including machine operators, supervision, programming, methods and processing, and tooling. We get tremendous input from the operators just by including them in the overall process. We feel it is a mistake to set a very reasonable speed and feed and then lock out the feed rate override.

Mr. Wolfe: Another key to success is the organization of a complete package before the job goes on the machine tool. We establish all tooling, processing, feeds, speeds, and all of the necessary documentation.

Mr. Harling: We have found that the best programmers have a good shop background. He is at the machine during the initial run, which keeps him in touch with the real world. Thus, he can actually see how his program functions. This gives him invaluable insight into better programming steps that he can later employ.

What do you foresee for the future?

Mr. Wolfe: Unquestionably, it will be computer numerical control and the use of the computer for more shop functions such as scheduling, processing, inspecting, material handling, and many others. **MMS**

Fig. 5—A belt-driven right-angle milling unit was attached to an NC machine spindle for generating close-tolerance slots in the workpiece, which is mounted on a vertical index table.

Fig. 6—More workpieces can be produced in an 8-hour shift by using multiple setups. Any given tool is used on all six workpieces before it is changed. The operator starts to change workpieces while the last tool is machining the last part.

Reprinted from Production Engineering, April 1979.
© Penton/IPC, 1979

Building Up Without Falling Down

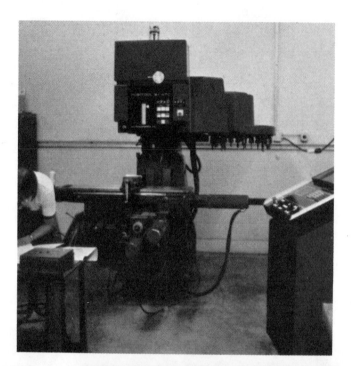

Can you really modernize a factory or boost its capacity without missing a productive heartbeat? Only sometimes, it seems. And there are lessons for all production engineers from the experiences of those who've been there.

By DONALD E. HEGLAND
Associate Editor

Machine tools are often at the heart of an expansion or modernization program. Autonumerics installed one of their own products to boost their machine shop's productivity; above. And Allis-Chalmers is grouping machining centers into high-production machining complexes. At right is one of a pair installed face-to-face for service by a single operator.

"We found a good opportunity to modernize because of advances in manufacturing technology," says Odded V. Leopando, manager of manufacturing engineering at Allis-Chalmers Corp. Engine Div.'s plant in Harvey, Ill. The engine plant manufactures diesel engines for the commercial market as well as for Allis-Chalmers. Three years ago, when the Engine Div. began to expand service to the commercial market, "The Division saw that the time was ripe for upgrading to meet a surging demand for diesel engines," says general plant manager Kenneth L. Schaufelberger.

Today, the engine plant is two years into a five-year, $20 million program which began with modernization and technology improvement and grew to include production capacity expansion. Summarizing the program, Schaufelberger explains, "Our first objective was the bottleneck operations in the shop. For example, our total unit capacity was restricted by insufficient capacity for crankshafts. Once the bottlenecks were identified, and plans laid to resolve them, our next objectives were cost reduction and quality improvement. And a primary concern throughout was making sure that we could produce new products using the best possible equipment. I think this is a fairly common way to set priorities and allocate capital."

Although the program includes improvement projects ranging from plant lighting through material handling, the major thrust is toward improved machining facilities. And the approach taken is mainly to consolidate operations on multifunction machines—especially CNC machining centers—wherever possible. Seven new chip-makers—four CNC machining centers and three CNC chuckers—are already in, nine more—all CNC—are on order, and an undisclosed number are in the offing.

"From the beginning," says Leopando, "the mandate has been to carry out the program without interrupting production continuity." The strategy they've used to accomplish this is twofold: first, the usual tactic of building a bank of parts to carry them through a machine debugging, and second, a clever approach to subcontracting.

"When a new machine, especially a CNC machining center, is ordered," explains Leopando, "we seek out a job shop operating the same machine. Then we subcontract with them to design and build fixtures—which we pay for—and run production lots of parts on *the same machine we are buying.*" The engine plant retains ownership of the fixtures, and when their own machine arrives, brings the fixtures back in-house, proves out their machine, and starts running production.

"We've done that extensively with the last six CNC machining centers we bought," says Schaufelberger, "and it has several advantages. You have the part fixtured c rrectly right at the beginning, and the tapes are programmed and proven out. There may be some optimization needed, but you're pretty sure you can produce a quality part right away with minimum problems." This stratagem not only keeps production going, it also starts paying off some of the cost-reduction and part quality improvement benefits before the new machine is even built. And it eliminates the on-line parts shortages and out-of-routing

"For our kind of manufacturing, the technology of tomorrow is going to lie in head-changers. . . . They provide a lot of flexibility and also the capacity to increase your unit build easily."
K. L. Schaufelberger, general plant manager, Engine Div., Allis-Chalmers

An NC crankshaft balancer has turned out to be one of the most rewarding projects in Allis-Chalmers' modernization/expansion program. With 17 different crankshafts in production, in relatively small lots, the flexibility and quick changeover provided by NC is a big plus for profitable production. And the engine test center is on next year's agenda. The 34 inefficient cells presently in use will be replaced by only 17 computerized test cells, for a net doubling of productivity.

"I cannot overemphasize the positive relationship between Allis-Chalmers' corporate Manufacturing Engineering Certification program and our modernization/expansion program. Our certified status made it much easier for us to plan our program and manage the execution of those plans."

Odded V. Leopando, manager, manufacturing engineering, Engine Div., Allis-Chalmers

situations usually associated with the transition from one machining process to another that is totally different. "After all," comments Leopando, "there's always a way to make a part, but in these situations it may not be a profitable way."

In a program like the engine plant modernization/expansion, stretching as it does over several years, manufacturing engineers have to accept as a fact of life that machines will be relocated as the program unfolds. For example, Leopando notes that they are relocating aisle lines to ultimately improve workflow. "So today, you may see a new machine encroaching on an aisle," he explains, "but two years from now either the aisle or the machine will be moved, and this temporary condition will be resolved."

Commenting on the program's benefits thus far, Leopando notes, "The machining centers, for example, have improved our workflow greatly. Foremen often used to be little more than glorified stock chasers, spending all their time tracking parts. And our work-in-process inventory has probably been cut by 75%; we just don't have it anymore, especially in detail parts like elbows, manifolds, and brackets."

Looking at future trends, Schaufelberger comments that the engine plant is essentially a very large job shop, and "For our kind of manufacturing, the technology of tomorrow is going to lie in headchanger equipment as opposed to the high-volume transfer line. I think you'll see a lot of headchanging machines installed in intermediate-volume manufacturing facilities in the next decade, because they provide a lot of flexibility and also the capacity to increase your unit build on a given model simply by adding another head-changer to the line."

MEC is the way to go

The engine plant's modernization/expansion program is inextricably intertwined with Allis-Chalmers Corp.'s formal Manufacturing Engineering Certification (MEC) program. The objective of MEC is to improve a plant's manufacturing engineering organization, systems, and performance, with an ultimate goal of improving profit through most efficient use of capital investment and employee resources. MEC separates the manufacturing engineering function into these ten discrete subfunctions, each of which can be measured independently, at least to a first approximation:

- Manufacturing engineering administration.
- Process engineering.
- Industrial engineering.
- Plant engineering.
- Tool engineering.
- Advance manufacturing engineering.
- Capital resource management.
- Material handling and plant layout.
- Cost reduction.
- Manufacturing systems.

The engine plant achieved this coveted status last year, the 19th of A-C's 34 plants to do so. And Leopando credits the disciplines required to attain and maintain MEC for much of their success. "Our modernization/capacity expansion program touches on about five of the ten subfunctions in MEC," he explains, "including those directly related to our capability to make parts to the best quality and highest profitability, and to our capability to increase production when the market demands."

"I think that MEC is a good management tool," Leopando continues. "The MEC manuals provide guidelines for all my man-

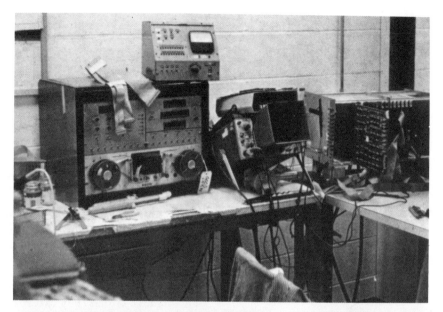

A major thrust of Autonumerics' expansion program was to give people more room in which to work. The number of circuit board testing stations, which use table-top minicomputers, was doubled and enough space remains to double it again. The primary objective in final assembly was to be able to work on more machines at one time, and surrounding the stock room with final assembly stations also simplified parts flow. A few areas have excess capacity; the wave soldering machine—used to solder all the components on a board simultaneously—only runs once a week.

agers that allow us to use our time much more efficiently. Without MEC, it would have been impossible—with the number of people I have—to keep production increasing and also take on the modernization."

"So I cannot overemphasize," he stresses, "the positive relationship between MEC and our modernization/capacity expansion program. Our certified status made it much easier for us to plan our program and manage the execution of those plans."

Meanwhile, out on Long Island . . .

"The basic driving force behind our move into a bigger building was to increase our production capacity to meet the projected increase in sales," says Autonumerics' chairman of the board, Warren E. Ponemon. After five years in their previous building, the Hauppauge, N.Y. machine tool builder's shops were so congested with equipment and work-in-process that people could no longer work efficiently, and stock and finished goods were overflowing into warehouses.

The 85-employee firm is basically a builder of sophisticated CNC controls, fabricator of small parts and complex toolchangers, and finally a machine assembler. They don't "build the basic iron," as Ponemon puts it; they buy the basic machine and create a CNC machining center by adding closed-loop ball-screw servo drives, variable-speed spindle drives, toolchangers, and the complete CNC control package.

So their primary thrust in moving into a 2½ times bigger building was to gain more final assembly room. And a secondary thrust was to gain space to expand their machining, testing, and customer support activities. For example, they installed one of their own products—a CNC machining center with all the options—in their machine shop to boost output of the myriad small parts that go into toolchangers and ball-screw installations.

In their previously cramped quarters, even a mundane task like painting parts, control panels, and finished machines presented a major problem, so they added a paint room. They also installed a hot room, in which they can run a fully computerized test on the completed control system while it is held at 125F for 48 hr. Even the number of bench-top minicomputer test stations for testing circuit boards was doubled. And finally, there's space for a formal classroom used several times weekly to conduct programming classes for customers.

Being by his own admission an incurable optimist, Ponemon recalls, "I didn't expect to lose any significant amount of production in this move." After all, they were only moving a few hundred yards down the street. So he wasn't faced with one of management's darkest fears—losing the workforce and having to train new people. And because they don't build the basic iron, there weren't any massive planer mills and things like that to move. "So our strategy," he continues, "was to stage the move—department by department—over a month's time."

Following the common tactic of building a bank of parts to cover each department's moving day, they worked a lot of overtime to stockpile circuit boards, complete control systems, control-to-machine wiring harnesses, and detail machine parts. And while this was going on, the new building was being stocked with common hardware supplies and basic machines from their supplier.

The planned sequence, explains Ponemon, was to move the electronics builders in first, while everyone else stayed behind to finish the last batch of machines that would go through the old shops. Then other departments would move as their jobs on these machines were completed—the machine shop first, then the wiring installers and machine assemblers, and finally the testing department. And so the total lost time should amount to about five minutes, right? "Wrong," exclaims Ponemon, "it just doesn't happen." The move took about two months, from start to finish, and he estimates that one month was essentially a dead loss to production.

As is typical, the lost time isn't attributable to a single major cause. But looking back, Ponemon can point to several culprits, himself included. Explaining that he wears the production engineering manager's hat as well as the board chairman's, he confesses, "I'm a typical entrepreneur, with all their drawbacks, specifically wanting to do everything." Consequently, he admits that one of their moving problems was that he was spread too thin over the project, in spite of the fact that, "My people keep reminding me that they can live without me."

In fact—excluding all the usual new-building glitches like having the building done on time (it wasn't) and getting all the right utility services in the right places (they weren't)—most of the delays were related to people in one way or another. For example, the machining center in their machine shop was intended to be used as a machining center should be used—for single-setup completion of a part. But for all that he is a

"An ideal way to accomplish a major capacity expansion would be to duplicate the facility, scaled up of course, while running full speed in the old one. Then move in and sell off the old equipment. In retrospect, I wish we could have.**"**

Warren E. Ponemon, chairman of the board, Autonumerics

CNC machine *builder,* Ponemon had all the usual problems getting shop people with a lifetime of experience on manual machines to *use* the CNC. Also, he adds, "It's incredible how many tools that people use daily are shared with another worker. As parts of the workforce moved, some of these tools went down the street too. So people went to get them, and human nature being what it is, took extra time to explore the new building, chat with friends, etc. No big individual thing but it all adds up."

"I think," he continues, "an ideal way to accomplish a major capacity expansion would be to duplicate the facility, scaled up of course, while running full speed in the old one. Don't strip the one to make up the other. Then move in all at once and sell off the old equipment. That's probably the way to do do it, if you aren't pressed for cash, which a small company always is. And in retrospect, I wish we could have."

Reprinted from Machinery and Production Engineering, December 7, 1977

Trochoidal oil-pump parts machined under NC

Development of successful approximations to true trochoidal curves has enabled NC programmes to be prepared manually for producing satisfactory pump components on a Gorton Tapemaster vertical milling machine

F J Robinson and W Whiting,
Lakehead University, Canada

As part of a continuing research programme in the field of trochoidal machines, a small reversible rotary pump was needed to serve as the lubricator on a prototype steam expander. Earlier theoretical studies had shown that the equations used to define the rotor and stator profiles for the expander, could also be applied to generate the rotating elements of a pump having the desired characteristics. In an article in *Machinery*, 6 November 1974, the authors described the NC machining of trochoidal profiles. In that article, reference was made to *Wankel RC Engine*, by R F Ansdale, for details of trochoidal analysis. Another paper by the author (ASME 75-WA/DE-10) gave details of suitable trochoidal pump profiles.

From the above-mentioned articles, the complexity of the defining equations for modified epitrochoidal curves and their envelopes will be appreciated. The articles also permit ready understanding of the work necessary to generate the associated control tapes for NC machine tools. The APT language was employed to produce the control tapes to machine the RL-I expander rotor and stator profiles, which necessitated the use of a large digital computer. A similar procedure may be employed to make the rotating elements of the required oil pump, but a simpler means of programming was investigated and developed.

Profiles for the oil pump

The design chosen for the oil pump is of the well-established Gerotor type, complete with an eccentric reversing ring. Preliminary calculations indicated that a suitable unit would result from the following trochoidal parameters: eccentricity $e = 2.5$ mm, generating radius $R = 25.0$ mm, winding number $Z = 5$, and constant difference increment $c = -13$ mm. When such a large constant difference increment is employed in a trochoid, a very large proportion of the profile takes the form of circular arcs of radius c. In the case of the profiles here considered, this proportion is in excess of 80 per cent of the total. Furthermore, the portion of the profile which makes the most important part of the engagement is the circular arc section.

Diagram (a), Fig 1, indicates the profile of the inner rotor of the oil pump in its final approximated form, also with the cutter path specified for machining the form on a vertical-type machine with a 30·48-mm diameter end milling cutter. Diagram (b), Fig 1, shows the associated outer rotor element for the pump, and a broken line indicates the path for a 4·7625-mm diameter cutter.

In the course of the theoretical analysis, a major constraint as regards the profile that could be obtained was the ability to machine the outer rotor element. This factor was controlled by a decision to use a 4·7625-mm (3/16-in) cutter and with this constraint and the other chosen parameters, the profile shown in diagram (b), Fig 1, was established. With this profile the circular arc sections, which form most of the curve, have a radius of 13 mm ($\frac{1}{2}$ in). Positions for the centres of these arcs were established by simple geometry. Portions of smaller curvature in the profile, however, ought not to be circular arcs since they are constant difference curves to the basic trochoid. It may be seen that with the approximated profile developed, these small-curvature portions are formed by the cutter, and consequently also are circular arcs. The difference is so small – less than 0·025 mm and well within machining tolerances – that the resulting curve, made up completely of circular arcs, was considered satisfactory.

Greater difficulty was experienced with the inner-rotor profile. An approximate design, which also made use of the same circular arcs positioned by simple geometry, proved to depart considerably from the true trochoidal profile. A rotor of this form was, in fact, produced and tested, but it required a considerable hand fitting before it would run in the outer element. Even then it clearly did not fit the mating profile well. Under these conditions, rather poor pump performance was recorded.

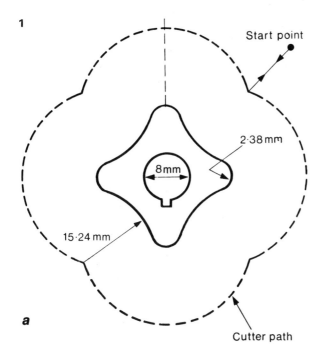

1

Start point

2·38 mm

8mm

15·24 mm

a

Cutter path

Start point

Cutter path

b

To overcome the above mentioned problems, a computer was employed to compare the true trochoidal curve and the circular-arc approximation. This work revealed interference between the two elements in excess of 0·16 mm, in the regions where the two arcs merge. It was also clear that an extremely close approximation to the true profile could be achieved by increasing the larger arc radius from 13 to 15·24 mm, also including some straight-line sections where the two arcs meet.

Fig 2 is a plot of one quarter of the inner rotor profile, showing the true trochoidal form and the final approximated form. From this figure it may be seen that there is no longer interference between the inner and outer rotor profiles, and the error is less than 0·1 mm. Furthermore, this error occurs in a region that will not adversely affect pump performance. The cutter path shown by the broken line in diagram (a), Fig 1, is for machining the final approximated profile by means of a special cutter, and the NC programme is only slightly more complicated should it be required to generate this profile with a standard end mill.

Machining the oil pump rotor elements
The vertical milling machine on which profiles were machined under NC on the inner and outer rotor elements was a Gorton Tapemaster, equipped with a Bunker-Ramo 3100 control system. This machine has some on-line computing facility, and is capable of machining circular arcs.

Programming may be carried out manually, with the aid of a Friden Flexowriter, and the word address type format is employed for the machine. Normally, when profiling operations are carried out on this machine, the number of information blocks is so large, having been generated by APT program-

1 *Inner and outer rotor elements for the trochoidal oil pump designed and made at Lakehead University. Diagram (a) shows the 'approximated' inner rotor and cutter path, and at (b) is seen the outer rotor element and cutter path*

2 *One quarter of the profile for the inner rotor*

element of the oil pump. Computer analysis was employed to establish the discrepancy between the approximate profile and the true trochoidal curve

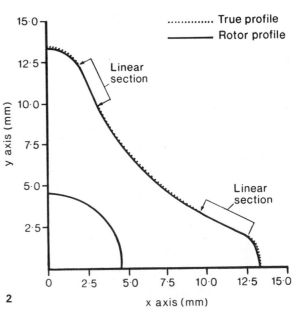

············· True profile
———— Rotor profile

Linear section

Linear section

y axis (mm)

x axis (mm)

2

Trochoidal oil pump parts

ming on a large computer, that automatic tape preparation is necessary. When a profile comprises merely straight lines and circular arcs, however, a control tape is far shorter. For the profiles described, the machining operations are expressed in 15 blocks for the outer rotor and 27 blocks for the inner element.

A simple fixture was produced to enable the workpieces to be mounted on the machine table, and this unit is illustrated in Fig 3. Blanks of the appropriate dimensions were machined from phosphor-bronze bar stock on a centre lathe and when mounted in the jig, the milling machine spindle was carefully centralised above the workpiece using a Heidenhain optical alignment unit mounted in the taper bore at the spindle nose.

In the case of the outer rotor member which may be seen seated in a bore towards the left-hand end of the fixture plate in Fig 3, the blank is circular and is machined to fit this seating. A central bore in the blank is slightly smaller in diameter than the minor diameter of the finished profile. A small keyway is provided in the periphery to prevent rotation of the blank during machining, and the blank is held in the recess by three screw clamps. Machining the profile in this blank entailed roughing passes, using a 3·175-mm ($\frac{1}{8}$-in) cutter at progressive depth increments of 1 mm until the rough profile had been cut through the depth of the blank. A single finishing pass was then taken at full depth with a 4·7625 mm ($\frac{3}{16}$-in) end mill. It was found that this procedure resulted in some chatter marks when the smaller curvatures were being machined. Since it had been noted that during the roughing passes the cutter appeared to pick up the profile extremely

3

4

3 *Fixture for machining operations on both inner and outer rotor blanks on a Gorton Tapemaster NC vertical milling machine. A machined outer rotor element may be seen in the appropriate seating towards the left-hand end of the unit*

4 *Components of the complete oil pump test unit. Two inner rotor elements are shown side by side. That in the foreground was machined to the approximate profile, and the other was produced from a much longer control tape to the true trochoidal curve*

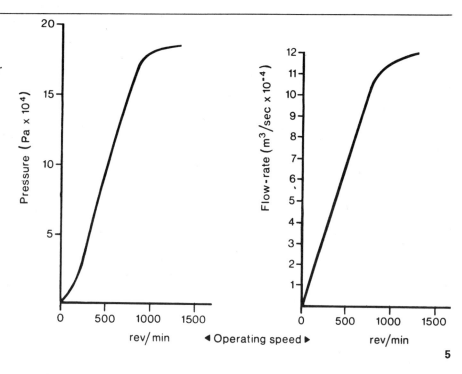

5 *Pressure and flow rate curves for the test pump unit equipped with the 'approximated' inner rotor element*

◄ Operating speed ►

5

accurately, the finishing stage was successfully carried out on a second blank by three successive passes to cover the entire surface width. Spindle speed employed for both cutters was 1960 rev/min, and the feed rate was manually controlled to obtain a satisfactory finish. No coolant or lubricant was used.

Blanks for the inner rotor were also machined from phosphor-bronze. Each was provided with a central bore of 8 mm ($\frac{5}{16}$ in) diameter and the keyway necessary for subsequently mounting on a pump shaft. On the fixture a silver-steel spigot was provided to locate the inner rotor blank, and use was made of the previously-mentioned keyway to prevent the workpiece from turning during machining. Centralising was again carried out with the aid of Heidenhain optical alignment unit, and roughing cuts were made with a 25-mm (1-in) diameter end mill at a spindle speed of 250 rev/min. The finishing pass was made at full depth with the special cutter, and also in three steps. As with the outer rotor slight chattering was experienced at full depth when machining the concave sections and this effect was eliminated by finish machining in three steps.

The complete oil pump

In order to test the profiles produced in this way, an oil pump test unit was designed and made. This unit incorporates a reversing ring, and is of conventional design. The components were produced by normal lathe and milling operations, and they are seen in Fig 4.

In this figure two inner rotor elements are shown, one of which was machined to the true trochoidal profile under the control of an NC tape prepared by APT programming on an IBM 360/50 computer. This control tape comprises more than 200 information blocks, and the component produced from it is so close to the approximated rotor, that only very careful measurement will detect the difference. Both rotors will run smoothly in the test pump, and the performance characteristics are very closely similar.

Several hours of test running have been conducted with the pump, using SAE30 motor oil, at shaft speeds between zero and 2000 rev/min. The unit reverses as required, but cavitates badly at about 800 rev/min due to the design of the intake port in the test pump. The unit in Fig 4 is 50 mm in diameter, and output curves for the pump equipped with the 'approximated' rotor are shown in Fig 5. ⊙

CHAPTER 2

PRACTICAL CASES

Reprinted from Machinery and Production Engineering, April 12, 1978

NC wins nearly all ways round

There are many ways to justify the purchase of NC machine tools these days. One fairly common reason is that NC can compensate for a lack of skilled operators. But even if there's no shortage of machining skills, NC can still pay off handsomely, as Arthur Astrop reports.

Northern Ireland is one of the few areas of the United Kingdom which does not lack skilled labour. On the face of it, therefore, one of the principal justifications for installing NC equipment would seem to be missing. Nevertheless, in the past five years the Sirocco Works of Davidson & Co Ltd, Belfast ('phone 0232 57251), has invested heavily in NC machines. 'Our need for NC,' says John Matier, general works manager, 'is not to make up for lack of skilled men but to augment those we have and above all to increase productivity per man.'

The extent to which Davidson has succeeded in this aim can be measured as follows. In 1972, the company had a turnover of £3 million and an annual output/man

valued at £3000. Last year, turnover was in double-figure millions, and output/man, £13 000. Despite the effects of inflation, the company is still confident that in real terms an increase in output/head of 100 per cent has been achieved.

The capital investment programme since Davidson joined the South African Abercorn Group in 1972 has been impressive, and is still continuing. Up to the end of 1977, £5 million had been invested in buildings, plant and equipment, and the programme calls for a further £1 million this year, and possibly next.

Since 1972, the company has been engaged in turning itself round from a deep-rooted traditionalism, accumulated over a 100-year period, to late 20th-century practice. The Belfast plant has many lofty large-area shops of an earlier age. It also has an inheritance of sound engineering practice which produces industrial fans that have a world-wide reputation for quality and reliability. In the past five years, Davidson has sought, with considerable success, to modernise the practice whilst retaining, and improving on, the quality of the product.

Sirocco works makes an exceptionally wide range of industrial fans. They cover small mobile units, which

NC machines that include a G & L-F horizontal borer with CNC have made considerable impact in the production of Sirocco bearings.

are made on a production basis, to massive power-station installations which may be a year or more in the making. Two areas of its manufacturing practice are particularly critical to the company's success, namely sheet and plate fabrication and high-accuracy milling and boring for its range of fan shaft bearings. It is in both these areas that the bulk of capital investment has been made.

As has so often been the case in recent years, sheet metal work in the Sirocco works has been revolutionised by the advent of the NC turret punch press. One Wiedemann machine now handles an estimated 70 per cent of all sheet items for components used in Sirocco 18- to 48-in fans of centrifugal and axial types. This machine, which has a General Electric controller, has a 30-tool magazine and its library of tapes now exceeds 600.

The Wiedemann, it is conceded, has virtually worked itself into an indispensable position as far as sheet metal work is concerned, and another machine of similar type may need to be installed to meet the de-

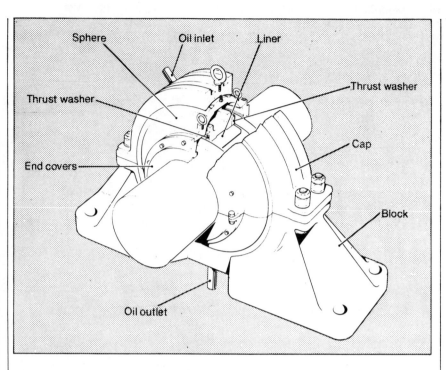

mands of increased business and to reduce the risks involved with the present dependence on one unit. The accuracy to which this machine will operate consistently has had a most marked effect on subsequent operations, including bending and folding of the blanks, and final assembly of casings.

Accuracy of blank developed forms is even more critical in plate work, particularly where rolling of profiled blanks into very large truncated cone forms is required. This work at present involves extensive marking out and flame-cutting from 1 to 1 drawings by magic-eye following. It is extremely difficult to achieve the required degree of accuracy on blank profile, and it is common practice to leave allowances in critical areas. These allowances must be flame-cut away by hand, when the plate is formed to its finished shape and the areas where adjustment is necessary can be identified.

Coming soon

New equipment shortly to be installed in Sirocco works will bring a major step forward in this area. Facilities for plasma-arc and gas cutting will be provided by a BOC 3-head machine with Kongsberg computer numerical control (CNC), which is to be installed at a cost of £150 000. This installation has a number of very interesting features. In addition to profiling plates at rates around 18 ft/min it will also automatically centre-punch all the points at which holes must subsequently be drilled.

Many of the plates need to have ribs and stiffeners welded into position after they have themselves been assembled to form large casings and similar structures. In the past, the positions of these auxiliary items have needed to be marked-out for the welders, an awkward and time-consuming procedure undertaken by hand. If the casing is very large, there are often problems of access. The BOC installation will have a powder-marking facility which will trace the positions for ribs and stiffeners at the same set-up used for plasma-arc profiling. The powder

makes a sharp colour contrast with the plate and will significantly ease the welders' problems at the later stages of assembling large structures. In addition, the installation will also have computer-assisted nesting routines, which will enable numbers of components to be profiled from standard plates with a minimum of wastage.

The single most important advantage which will be gained, however, is the increased accuracy to which profiles can be cut, under continuous-path CNC, compared with the present methods. It is anticipated that the need to leave excess metal on edge-joints, for trimming off after the plate has been formed to its finished shape, will be eliminated entirely.

This factor alone will make a major contribution to cutting production times. Davidson estimates that the build time for the stator for a large air preheater, which at present could be as long as 12 months, may well be reduced by as much as 30 per cent when the new BOC equipment is installed.

It is in bearing manufacture, however, that NC has made its greatest impact at Davidsons, and where the major investment has been placed. A new bearing shop has been constructed and laid out on most modern lines. Whilst Sirocco bearings are important components in the company's fans they are also units which stand in their own right, as independent self-contained items for a wide range of other applications.

Four basic types of Sirocco bear-

ings are made, namely self-aligning whitemetal lined (pressure-fed or ring-lubricated), tubular, and spherical roller. For pressure-fed whitemetal units, the normal range covers shafts from 90 to 335 mm diameter, and for ring-lubricated the sizes are 45 to 225 mm diameter. Housings are of the plummer-block type. Tubular Sirocco bearings incorporate deep groove ball and roller races, and anti-friction spherical bearings, which cover shaft sizes up to 170 mm diameter, have twin-row self-aligning roller bearings.

In the new bearing shop there are three large machines by Giddings & Lewis-Fraser, one with CNC, and a 10 HC numerically-controlled machining centre with automatic tool changing by Cincinnati Milacron. The principal production problems which have been tackled are the machining of the upper and lower halves of plummer-block castings (some of which are very large and heavy), machining internal and external spherical surfaces, and rough and finish boring of the bearing bores. Finish boring of the white metal is included in these operations.

The G & L-F and Cincinnati machines are substantial investments in themselves, but they have been backed up by a large number of specially-designed fixtures aimed at easing set-up and changeover, and thus at reducing floor-to-floor times even further. The latest G & L-F machine, for example, which has computer numerical control, is also provided with large double-sided fixtures

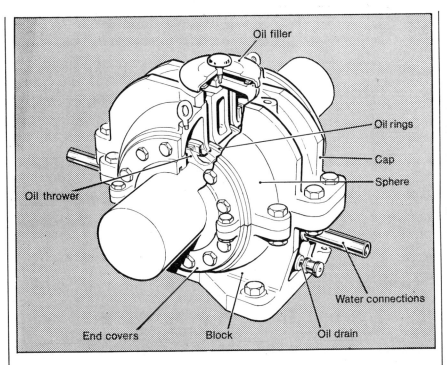

Oil filler

Oil rings

Cap

Sphere

Oil thrower

Water connections

End covers   Block   Oil drain

Two of the different types of bearing units made, with the help of numerical control, at the Sirocco works.

designed to be used on the indexing table.

One seating of the fixture holds a plummer-block casting for operations on one face, and the other seating — which can be unloaded and re-loaded during the machine cycle — holds an identical casting for operations on its opposite face. Easy location of castings in their seatings, and quick-acting clamping designed to avoid distortion, have probably brought down set-up times for these bulky castings to an irreducible minimum.

The Cincinnati machine is basically a standard 10 HC machining centre but has a special feature to suit Davidson's requirements. It has two indexing tables, but one is smaller than the other. The centres of the tables are off-set, by 100 mm, so that points on their peripheries adjacent to the spindle are in line. This arrangement minimises tool overhang when the smaller table is in use, and allows preset tools with a common basic 'reach' to be used for either table. Presetting is thus simplified.

The new bearing shop has its own presetting room, and many tools for the G & L-F and Cincinnati machining centres are dealt with by an Italian-built Speronic presetter which features direct-diameter reading for boring tools, thus avoiding the need for the operator to calculate radius values. Cutting edges can also be set for axial position on the same unit and at the same set-up.

Two more G & L-F machines in this shop, each with a Numeriset

positioning system, are used extensively for machining the internal and external spherical surfaces used in white-metalled Sirocco bearings of the self-aligning type. In this type of bearing, there is a cast-iron 'ball', with the white-metalled bore at its centre, which fits into a plummer-block split housing with a mating internal spherical surface.

Davidson has developed a range of special tooling to machine the spherical surfaces, which range in diameter from eight to 29 in. An average of $\frac{5}{16}$ in machining allowance is left on the spherical surfaces, internal and external alike, and is removed by a swivelling device actuated by axial movement of the machine spindle. A special faceplate is provided for attachment to the spindle, and the size range of spherical surfaces to be cut is covered by interchangeable tool holders of different lengths.

The use of machining centre practice, coupled with the specially-designed fixtures, has enabled the company to cut floor-to-floor times for many items used in the Sirocco bearing range by 50 per cent in some cases, and by as much as 80 per cent in others. An exceptional advantage offered by the latest G & F-L machine, which has CNC control, is the ability to programme for orbital milling.

The ends of the bearing housing castings have large-diameter, but

fairly narrow, annular facings which need to be machined. It is a simple matter to programme a facing cutter to move orbitally around each of these faces at a high feed rate, and the operations are completed in a significantly shorter cycle time than if multi-pass milling or radial-facing were to be used.

Davidson has found that for the grade of cast iron used in Sirocco bearings, Walter cutters perform best for the majority of operations, and Valenite tools for machining the internal and external spherical surfaces. NC is also used in the bearing shop for drilling operations, and a Richmond machine is installed for this purpose. For the NC machines currently employed by Davidson, which include a Wadkin with a retro-fitted Plessey controller, the company at present employs two full-time programmers, and two part-time. The part-time programmers are often students who are taken into the Sirocco works for a period as part of sandwich-course training.

For the future

The need for another full-time programmer is foreseen when the BOC plasma-arc cutter comes on stream, and especially when Davidson's future capital investment plans are realised. At least one more machining centre is planned for the near future, possibly two, and the company is evaluating NC turning.

In fact, Sirocco works has had a very large NC lathe (16 ft between centres) for some time, but the investment has not proved as successful as was hoped. To some extent it is the range of work which has proved the obstacle, in that the lathe is essentially for shaft turning and too many of the components for which it is suitable are needed only in 1-off quantities. As a result, the cost and effort of programming are disproportionately high, and it has become increasingly difficult to justify use of the machine.

The present investigation into NC turning, however, is concentrated on chucking work, and indications are that a powerful slant-bed turret-type machine will meet the needs. □

Reprinted from Modern Machine Shop, January 1978

Computerization Comes To Shearing

Optimum control of an expensive inventory, and greater manufacturing flexibility were the prime movers behind a computerized coil shearing system.

By KEN GETTELMAN, Associate Editor

The complete shearing system is nearly 205 feet in length, it has a price tag in the seven-figure range, it will very easily handle 25,000 tons of steel coil per year on a one-shift basis, it will handle coils ranging in width from 12 to 84 inches and in thickness from a 0.031-inch sheet to a 0.375-inch plate, it will cut lengths from a mere 9 inches to a hefty 70 feet, and its entire running and coordination is controlled by a computer that is programmed with a punched tape that is about two feet long. Obviously, it is not something the average shop would buy, yet it has completely changed the sheet-metal processing at Youngstown Steel Door Company in Youngstown, Ohio.

As one of the nation's leading fabricators of railroad car doors, sides, and other components, Youngstown is a large consumer of steel sheet and plate. In the past, the different size and gauge sheets were purchased in the precut and trimmed configuration, and placed in inventory. The costs of such precut sheet and plate, and of maintaining an adequate inventory were high. In fact, they were so high that Youngstown officials state the new Paxson shearing line will pay for itself in a very reasonable time.

The key factor is the ability to buy raw coil steel and shear either a portion or an entire coil as needed. The computer has made it possible for the line to shear either a specific number of sheets from a coil, a specified weight, or to process an entire

Overall view of NC controlled shearing line.

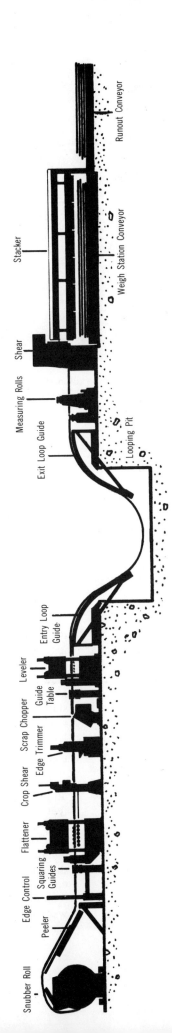

Diagram of complete shearing line.

coil and record both the number of sheets and the weight processed. It will compute and record both the amount of usable material and the reject scrap from a coil.

Railroad car building is not what it used to be. The day of the common box, tank, and gondola car has given way to highly engineered and specialized cars such as the Youngstown Bathtub for the high volume transportation and quick unloading of coal, the very large hy-cube box cars, and the balloon-type tank car. All are designed for maximum service and minimum weight. Thus, they are comprised of a number of different types and gauges of steel.

When Youngstown Steel Door Company officials began studying the problem, it became obvious that reduced material costs, inventory control and minimization were key factors to the economical production of a growing number of specialized components. As the product line grew and became more specialized so did the inventory of a large number of sheet metal components made from various types of steel.

Paxson Machine Company of Salem, Ohio became the prime contractor and did all the mechanical development, design and manufacturing work on the installation. The electrical and computer control portions, including programming, were done by Unico, Inc. of Franksville, Wisconsin. The computer itself is a Digital Equipment Corporation PDP-12.

Operation

As the operator faces the unit from his control station, the action begins at the left with coil loading. There is a turnstile which can hold up to four 50,000-pound coils. The turnstile is rotated so that the selected coil is brought into position to allow the coil car to get under the coil, lift it off the arm and shift it to the payoff reel of the line. It is the operator's responsibility to properly center the coil.

The power-controlled snubber and peeler are utilized by the operator to start the coil through the line. It goes through an edge control and squaring guide into a flattener, which flattens material that is thicker

After threading a coil of steel through the automatic shear line, operator John Zatchok inserts a two-foot length of punched tape, which contains information about shear length, number of pieces, and other instructions which the computer will use in controlling the line.

than ⅛ inch. The next station is a crop shear to square the leading end of the strip. An edge trimmer cuts off and chops up the rough edges of the strip to insure a square sheet at the final cut. A leveler flattens material that is thinner than ⅛ inch. An important part of the line is the loop pit, which allows the strip to accumulate for the time required for high speed shearing. The measuring rolls feed the shear, which cuts to the computer-programmed length. Stacking and runout are also computer controlled, placing either a predetermined number of sheets or weight in an individual stack for further processing.

Three Modes

With the assist and control of the computer, it is possible to run the line in three separate modes to meet any kind of shearing condition. The first is the "tight line mode" and is only used for the heaviest gauges of material or any steel that would take a "set" in the looping pit. In this mode, the strip is fed straight through and the computer synchronizes all feeding stations to get the proper length to the shear for cutoff.

The second feed method, and the one used for most shearing, is the

"loop line" which allows the strip to be constantly fed from the coil while the shear is run on a start/stop basis. The looping pit accumulates the difference between the constant feed from the coil and the start/stop action of the shear.

The third method is the loop line with a continuous shearing action. This particular mode is especially appropriate for very short lengths. Rather than constantly clutch and declutch the shear, the computer allows it to be run in a continuous mode with a controlled rate of strip feed to the shear so that the proper lengths are cut. The computer is constantly calculating and maintaining the correct feed rate to cut the right length.

Computer Output

Aside from operating the coil feed and shearing line at speeds up to 300 feet per minute with all functions fully coordinated, the computer also develops some very important management information. With each individual program the computer records and prints out the date, shift, operator number, the time the line went up, the type of steel, the coil number, the starting weight, the run time, the percentage efficiency during the up time, the line down time, how many pieces were produced, how many were within the plus or minus 1/64-inch tolerance, how many pounds of material were consumed, how much is left in the coil, how much scrap, and so on. Each day the purchasing department receives a complete rundown on consumption, and on the type and quantity of individual coils remaining in the raw material inventory. It is this type of information that management can use to better organize production schedules, purchasing policies, and the all-important inventory management. With steel averaging some $300 per ton, it doesn't take very many 25-ton coils to reveal the value of a well managed inventory.

No Magic

The computer is a very effective controlling and recording tool, but it is not an all-inclusive brain. It will take input from the many sensors placed along the line and balance feed, shearing, and so on; however, it is the responsibility of the operator to set up and control certain functions along the line. Neither the flattener nor the leveler is controlled by the computer. Here, operator experience has proved to be the best approach. He adjusts the roller spacings at the time of setup. It takes about six minutes to set up for a new coil of the same material and about 10 to 15 minutes if a different width and gauge of coil stock is scheduled for shearing. Nor will the computer think. The Industrial Engineering Department must still decide and program the size, type, and amount of sheets that will be needed. The computer is merely a tool.

New Youngstown Bathtub rotary-dump car for maximum loads and quick unloading of coal is one of many styles of railroad cars that make use of different types and gauges of steel. It carries 4½ tons more coal than conventional cars, yet weighs thousands of pounds less.

Perspective

Historically, most machine tools—whether the chip or forming type—have been justified on the basis of labor, tooling, or time savings. The computerized shearing line tackles these, plus others. As Amos Walker, Director of Industrial Engineering, and Martin Humphries, Assistant to Mr. Walker, points out, the justification results from the savings in material costs, scrap and lower inventories. In addition, the line has provided a much better manufacturing flexibility with shorter runs and better scheduling to meet market demand. A much better customer response has been a benefit that is very hard to quantify. With a more precisely sheared sheet, it is possible to control manufacturing better and to develop a higher quality product.

Since the first coil was sheared in December 1976, the Industrial Engineering Department has learned how to process the sheets with a greater degree of efficiency. The success with the coil line has also led to considerations for better modernization of the downstream processing of the cut sheets. An NC turret punch press is under consideration for those jobs that do not warrant expensive tooling and setup on a standard gap press.

According to Mr. Walker, "The heavy sheet and plate industries are moving from an era of brawn to brain." **MMS**

DNC experience leads to CNC

**Production tests with DNC led to specifications for
standard NC systems: a DNC-compatible CNC system that
accepts CL data for machine-independent part programming**

Manufacturing facilities often grow like Topsy, and uniform machine and control systems throughout a plant are a rarity. This has been particularly true for some of the pioneers in the application of numerical control because NC equipment has gone through several design-generation changes in a relatively short period of time.

Take the B-1 Div of Rockwell International in Los Angeles, for example. Among the 53 NC machines at the facility, some date back to 1958, literally the dawn of NC technology. Until recently, there were seven different control systems in use, but, with the various NC/machine combinations, 13 unique tape formats were required to prepare part programs for all machine tools, and it took 23 post-processors to suit source programs to the various machines.

The oldest NC is a GE/Concord control, a first-generation magnetic-tape/analog system installed on an opposed-head cavity mill as part of the initial Air Force NC-purchasing program. Although the vacuum-tube-type control is still operative and there is occasional work for the machine, Rockwell no longer has the programming facility to produce analog magnetic tapes.

Such system variety spawns many problems for a manufacturing operation. Large and costly spare-parts inventories are required; it is almost impossible to train maintenance personnel to handle every system; it is equally difficult to have a group of machine operators versed in the operation of different systems on similar machines; and, of course, part programming is vastly complicated by the large variety of tape formats required.

These were some of the underlying reasons leading to an ongoing modernization program that involves retro-

By George Schaffer, associate editor

Operator inserts flexible disk (floppy) into new breed of CNC system furnished by Vega to meet Rockwell specifications for standard NC. System accepts CL data

fitting 31 NC machines in the Structural Machining Center at the B-1 Div with identical CNC systems that meet standard NC-system specifications developed by Rockwell personnel. Once all of these new-generation CNC systems have been installed, all the variety-related problems will disappear, and, in addition, the shop will feature complete part/machine independence. Within the

limits of machine travel and capability, part programs will work on any CNC-equipped machine tool without requiring additional postprocessing to suit the program to a particular machine.

And, in terms of long-range development, the system as installed is also suitable for upgrading to DNC operation without significant additional hardware or software. Arthur Welch, CAD/CAM

DNC experience leads to CNC

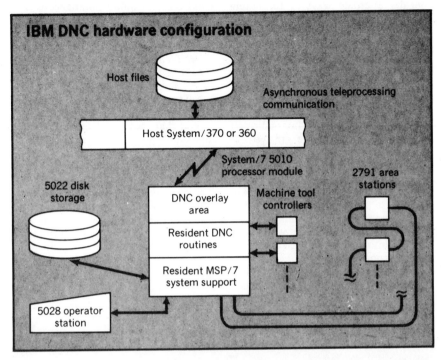

IBM DNC hardware configuration

Host files

Asynchronous teleprocessing communication

Host System/370 or 360

System/7 5010 processor module

5022 disk storage

DNC overlay area

Resident DNC routines

Resident MSP/7 system support

Machine tool controllers

2791 area stations

5028 operator station

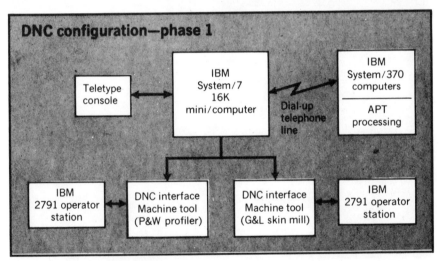

DNC configuration—phase 1

Teletype console

IBM System/7 16K mini/computer

Dial-up telephone line

IBM System/370 computers

APT processing

IBM 2791 operator station

DNC interface Machine tool (P&W profiler)

DNC interface Machine tool (G&L skin mill)

IBM 2791 operator station

Then, in early 1973, IBM was interested in testing and validating its System/7 minicomputer-based DNC approach in a production environment, and Welch was approached with a tempting proposition: IBM would supply the computers and software if B-1 Div would furnish the machine tools for a DNC test hookup. "That seemed like an excellent opportunity to gain some experience with DNC at very little cost, and we entered into a joint venture with IBM," Welch recounts. Here was a means of developing useful data for a subsequent decision on the acquisition of some type of automated machining system.

The IBM DNC system involved transmitting NC programs from a host computer, located at Rockwell's Western Computer Center at Downey, Calif, to a System/7 computer at B-1 Div. These programs, stored in the System/7, were then transmitted to selected machine tools as required. The DNC interface was directly behind the tape readers of the machine tools in the data link, permitting the machines to revert readily to independent tape-reader operation.

The DNC system provided by IBM for the test had the following hardware:
- IBM System/7 computer with 16k core and two disks
- IBM 5028 operator keyboard station
- IBM 2791 communications terminal (operator badge/card station)
- Machine-tool interface for a GE Mark Century 100M and a Bendix 1503 NC systems
- Dial-up telephone line with an acoustic coupler.

The binary-synchronous communications software installed and tested by IBM was modified by Rockwell to provide the necessary link with the Rockwell APT NC-programming system. The initial test phase involved two machine tools: a Giddings & Lewis skin mill and a Pratt & Whitney profiler. A subsequent production test phase involved five machines in the DNC loop (one G&L skin mill, two P&W profilers, one Kearney & Trecker profiler, one G&L five-axis profiler) and an additional five machines in a communications loop. The communications loop recorded productivity data for comparison with the five machines operated by DNC.

The production test lasted for 180 days and proved that the hardware and software were capable of operating the machines in a DNC mode and that the system could also monitor and store relevant production data. But the tests also provided these valuable insights into the DNC setup:
- To operate efficiently, the system requires the use of a dedicated high-

coordinator for B-1 Operations, puts it this way: "We have a CNC system that is totally DNC-compatible—without any of the drawbacks of DNC."

Here is how Welch compares CNC (computer numerical control) and DNC (direct numerical control):

DNC implies transmitting part-program outputs from a host computer or computers to a series of machine tools in a manufacturing-data link. Tape readers either are bypassed or do not exist at all. If the host computer goes down, all machines in the data link go down until the problem is resolved. In contrast, CNC merely replaces the existing NC system with a computer and software to control the motions of an individual machine. As each machine has its own minicomputer, only one machine is affected if trouble should occur.

Input to a CNC system is usually a form of punched or magnetic tape from a reader on the system, but, with DNC, the data are transmitted from the host computer directly to all the machine tools in the system over appropriate data lines.

Welch obviously shows a preference for CNC with its relative independence from central computer-related shutdowns, but DNC was also under consideration during the planning stages for the present modernization program. "We knew that the control systems on these machines had to be changed because they were getting old and we wanted to eliminate the many varieties of systems. At the same time we were looking at DNC possibilities. But we really didn't know which way to go," he says.

speed bisynchronous line instead of the dial-up telephone line to connect the System/7 to its host computer. This reduces the time needed to transmit lengthy part programs to the System/7.

■ The shop must follow its planned parts sequence or provide sufficient advance notice of a part substitution to achieve timely access to the DNC data.

■ The 2791 operator terminals are physically too large and complex to install on a traveling-gantry machine, but a remote stationary location would involve excessive operator time in running the machine.

■ System/7 disk-storage capacity needs to be expanded to store multiple jobs for five machine tools.

■ DNC-system support is required on all operating shifts.

■ An NC data-file system should be included to store and retrieve programs from the host computer, because the disk capacity at the System/7 is limited, and, when it is on-line to the host computer, all jobs are transmitted whether they are needed or not.

■ Full-time system-programmer support is necessary.

One thing was clear: An effective DNC system is dependent on custom-tailored hardware and software. Other conclusions: An operational and effective DNC system is expensive in startup costs, must be supported with software programmers and hardware maintenance, and must be backed by all levels of management. Even then, the Rockwell tests indicated that productivity increases are not automatic and projected savings are usually intangible.

It was quite clear that DNC could not be justified for the B-1 Operation, and so available conventional NC and CNC systems were reviewed to determine whether a "standard" control could be selected for all of the major profilers in the Structural Machining Center. The idea was to use standard hardware and identical input media for all controls and eliminate control problems like:

■ Old age—average control age 9.29 years.

■ Three types of tape required: Punched Mylar, reel-to-reel magnetic, cassette

■ 13 tape formats required

■ High maintenance downtime and costs

■ Tape-punch failures

■ Cassette-kit problems

■ Concord control obsolete

■ Reliability

■ Reprogramming costs for running the same parts on different machines.

"But all of the controls on the market at the time still made use of punched tape or cassettes and the traditional

postprocessors for each machine tool," says Welch. "In view of our goal of a standard control, it was desirable to consider alternate ways of formatting the data and selecting a better input medium." Therefore, a purchase specification was drawn up with standard control requirements included as mandatory items. Here are some of these items that were different from previous Rockwell specifications for conventional NC:

■ The NC system shall internalize all magnetic functions to the greatest extent possible.

■ The system shall be modular and expandable, so that only minor modifications to software and hardware are required to adapt the control to other types of machines (such as lathes or machining centers) and to integrate new features into the control (like DNC or adaptive control).

■ The system shall be a digital and/or analog input network of the lat-

est approved type and shall include a minicomputer or microcomputer in conjunction with all necessary equipment for controlling simultaneous motion and auxiliary operations of machines with as many axes of motion as specified and with operation at maximum feedrates.

■ Data-input device shall be a direct access, flexible-disk system.

■ The input-data format shall be Rockwell's CL (cutter-centerline) data format that provides a single, standard input to all control systems. All CL codes to conform to APT standards where possible, with documentation provided by Rockwell.

■ The seller shall provide system-maintenance programs to automatically diagnose the control system to the major-component level and identify faulty components with suitable indicators.

The most controversial items of the specification were the requirements for accepting CL data and the use of flexible

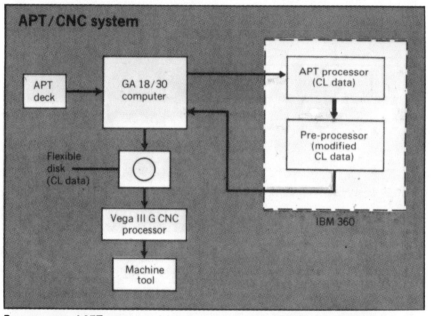

Preprocessor at APT processor (above) arranges data in IBM 3741 data format for writing. Disk-writing system (below) has dual flexible-disk read/write unit for disk copying The GA 18/30 computer is necessary for link with corporate computer

DNC experience leads to CNC

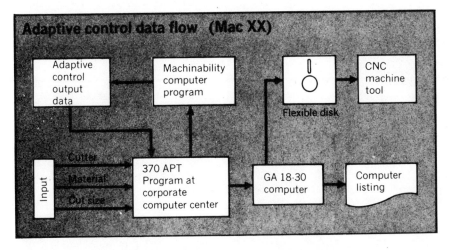

Adaptive control data flow (Mac XX)

Output of machinability processor is a set of machining conditions required by the MAC XX adaptive control. Output of flexible disk is series of APT statements

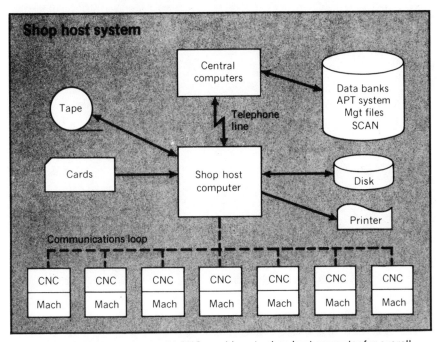

Shop host system

Future plans include linking the 31 CNC machines to shop host computer for overall management-data collection and monitoring, including sensor-based maintenance

ability to modify the software and adapt the controls to integrate other control features. A new Macotech computer-adaptive control has been interfaced to the software of the first of 15 III G CNCs to permit use of universal force codes acceptable to any machine tool in the system. Again, it is a matter of providing for complete part/machine independence within the system.

The Macotech MAC XX system is a further development of the MAC I adaptive-control system. MAC I achieves maximum chip load and maximum metal removal in milling operations by monitoring cutter deflection and spindle load and then controlling cutter force and spindle horsepower. A machinability computer program prescribes the safe load limits for cutters and also computes the spindle speed and maximum feedrates to be used at any given time. This information—the force codes—is automatically put on the part program when the source program is processed through a system like APT.

The MAC I computer program is actually a software preprocessor using information like workpiece material, cutter size, and maximum stock that is encountered—information supplied by the part programmer or process planner. But, with MAC I, the characteristics of each different machine tool—spindle stiffness, maximum allowable spindle load, and horsepower capacity—are permanently stored in the machinability computer program and used in part-

program preparation as a particular machine is specified.

With MAC XX, a softwired program has been developed that modifies adaptive tape instructions to match machine characteristics, and this program can be added to the CNC software controlling the particular machine. With this arrangement, only one part program containing adaptive-control instructions is needed for optimum productivity on any machine equipped with a MAC XX control unit.

Integration of the MAC XX adaptive-control system into the Vega CNCs involves these elements:

■ A software preprocessor similar to the MAC I works in conjunction with the APT processor at Downey to put universal adaptive-control statements on the part program. These statements accompany the CL data going on the flexible disks for CNC data input.

■ A software adaptive-control operating system is integrated with the Vega CNC at the machine tool. This routine modifies the adaptive-control statements to suit the strength and power of the particular machine. The Vega CRT editing feature can be used to edit these statements for fine tuning or extraordinary conditions during proveout.

■ Finally, there is a hardwired control section mounted inside the Vega control cabinet that processes force- and horsepower-sensor signals to provide proportional voltage signals.

What about the future? Each of the Vega CNCs also has the ability to monitor 256 dc inputs and outputs, in addition to normal CNC functions. Future plans include using this capability to link the 31 CNC systems to a shop host computer to provide a comprehensive monitoring and management-data-collection system. Also to be included is a sensor-based maintenance system to monitor machine tools in the system.

And, because the CL-data format is computer compatible (CL data could be fed directly from a computer to the CNC system without need for EIA or ASCII conversion), DNC is just a matter of bypassing the flexible disks. "Actually, the benefits from a comprehensive management-data-collection system are greater than those from DNC, but we didn't want to rule out the possibility of driving the machines from data coming directly from a host computer," says Welch.

In fact, Welch feels that the current DNC philosophy has many drawbacks. "What we have done at the B-1 Div is take the good features of DNC and bypass the problem areas." Not a bad approach when it comes to nibbling away at the CAM elephant. ■

Reprinted from Machinery and Production Engineering, February 22, 1978

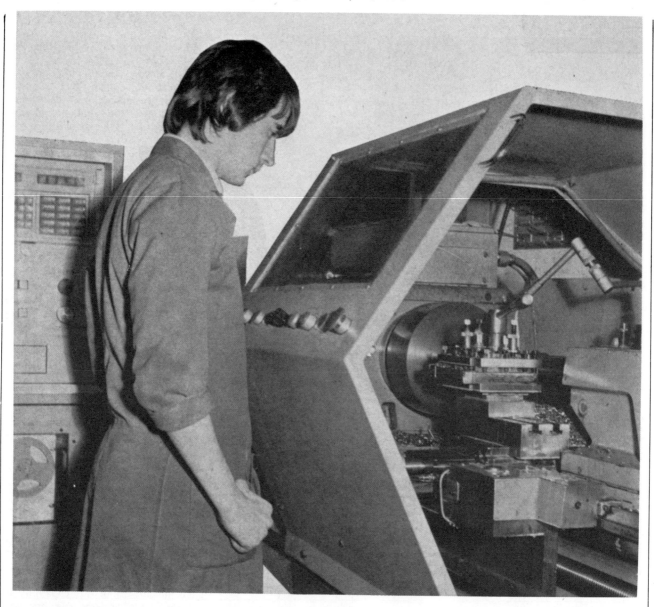

Are the facilities generally offered by CNC lathes too extensive and the costs too high for the smaller type of firm? One company in the South thought they were, and has introduced a low-cost model built up from standard products. Bob Sansom gives the details.

Cutting the cost of CNC down to size

Michael Young Engineering Ltd, Fareham, became alarmed when the scrap rate reached a level of 28 per cent during production of a particular type of component. At that time, production was carried out on a manually-operated lathe and occupied 6 h/part. Enquiries indicated that similar parts were being produced by another firm in a cycle time of only 4 min, although in that instance, a high-cost CNC lathe was employed.

The capital investment that would have been needed to provide similar facilities at the Fareham works proved to be unacceptably high. Moreover, when the specification of the CNC lathe was compared with design requirements of the product, it was considered that the production methods had been over-engineered by the second company.

Other CNC lathes were investigated, but all provided facilities that were considered too extensive, and none was available at a cost which Michael Young Engineering was able to justify. For this reason, the company decided to build a low-cost CNC lathe to suit their requirements, and the resulting machine is based on a Colchester Triumph 2000 centre lathe and the Posidata CNC 2800 system. CNC lathes of this type have now been made available to other firms, and they are marketed by a newly-formed associate company, Timatic Systems Ltd ('phone 03292 82751). Prices range from £20 000 to £26 000 depending on tooling equipment that is specified, and delivery in 4 to 4½ months is being quoted initially. Stepping motors normally provide drives

posts, and operation of the tailstock are performed manually.

Work involved in converting the Colchester lathe for CNC operation includes principally, replacing the saddle leadscrew and cross slide screw with circulating-ball screws to ensure freedom from backlash, and fitting of stepping motors. In addition, the feed gearbox and shaft have been removed,

time for producing the component mentioned earlier to be reduced from 6 h to 19 min, and scrap has been virtually eliminated. A drive shaft for a water pump which is also produced on the lathe has three diameter steps, one of which is threaded, and the cycle time has been reduced from 20 min to 4 min. Work involved in producing piston rods for air cylinders includes turning three diameters, and screwcutting two of the resulting steps, with threads extending close to the shoulder face in one instance. These operations are now being completed in 27 s, whereas 2 min were required formerly.

on the x and z axes, but dc motors can be supplied at an extra cost of £4000 to £5000, and such units would provide for contour machining.

In developing the lathe, the company has been able to draw on experience in electronics, computer operation and retrofitting of NC equipment. As regards retrofitting, the company applied an earlier Posidata hard-wired NC system to a well-used milling machine some time ago, and the combination has since been in continuous operation in the works on production duties.

The company has adopted the principle that the CNC system should take over only those functions in which operator skill is required in turning. Whereas there is provision for single-point screwcutting as well as turning and surfacing operations under the control of the Posidata system, auxiliary functions such as changing of spindle speeds, indexing of tool-

Two views of the low-cost Colchester/ Posidata CNC lathe developed by Michael Young at its Fareham works.

and the saddle apron has been replaced with a plate for mounting the nut associated with the x axis ball screw. A Dickson 4-way toolpost at the front of the cross slide can be indexed in 15° steps and is located by means of a Hirth coupling. A second Dickson toolpost can be mounted at the rear of the cross slide. The tailstock is fitted with an SKF live centre, and a one-shot lubrication system is provided. Duplicate controls for starting, stopping and reversing the headstock spindle, slide hold and release, and jog operation, are included in the Posidata console and on a sliding-type guard that normally encloses the machining area.

The Colchester/Posidata lathe that has been brought into operation at the Fareham factory has enabled the cycle

Keyboard programming

Programming for fresh work is carried out by the operator with the keyboard incorporated in the Posidata system. After a workpiece blank has been set up in the chuck or between centres, the toolpost is brought to a convenient zero point in relation to the headstock and the work axis. Cutting paths for producing the component are then entered into the computer memory, and after a trial cut has been taken on the part, any correction that may be required is applied to the programme by means of editing facilities.

Input data can be applied to the Posidata equipment in inch or metric units, feed rates can be varied in 0·004-in steps in the range from 0 to 120 in/min, and screw pitches from 10 to 50 threads/in can be obtained. There is provision for applying a total of 24 tool offsets for turning diameter and length, also tip radius.

After a batch of work has been completed, a reference tape is prepared from optimised data stored in the memory. Particulars of the workpiece, tooling, and zero positions are entered into the heading of the tape with the aid of a Flexowriter. Michael Young Engineering will consider converting lathes by other builders for CNC operation, and facilities for constant cutting speed and tool changing can then be provided if required. □

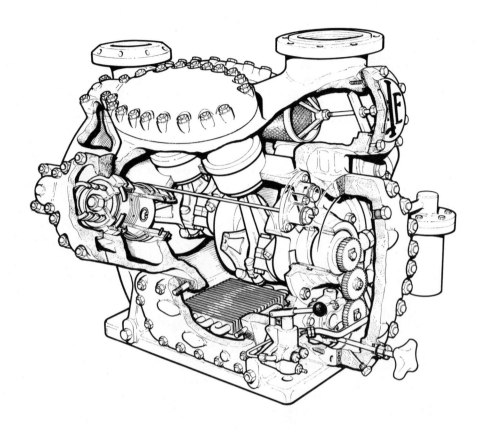

Reprinted from Machinery and Production Engineering, January 14, 1976

NC machining of large castings

Hall-Thermotank Products Ltd use gauging fixtures to check and prepare large castings for machining on machining centres, on which some of the larger tools employed have aluminium-alloy bodies to reduce the weight

B C Kellock, Associate Editor

The company known as J & E Hall has been associated with refrigeration since the early days of carbon dioxide equipment, which was introduced towards the end of the last century as a replacement for the earlier cold-air machines. Interest in food

preservation extends back even further to 1813, when two pioneers of the company started the world's first canning factory. Today, the name J & E Hall is used as a product trade name within the recently-formed company Hall-Thermotank Products Ltd, a member of the Hall-Thermotank Group, an international organisation concerned with production and marketing of a wide range of air conditioning and refrigeration plant for use on land and at sea. It has been estimated that 53 per cent of the world's marine refrigerated cargo-carrying capacity registered at Lloyd's is cooled by refrigeration equipment designed and built by J & E Hall.

Most of the refrigeration plant built by the company employs the principle of vapour compression for the basic heat extraction process, and an essential component of such systems is the compressor. In consequence, compressors form a very large part of the overall output of the Hall-Thermotank Products factory at Dartford, Kent, where some 1200 people are employed. These compressors are of the reciprocating-piston type, and the design is such that all forms are able to operate in a wide range of working conditions both with regard to the refrigerant used and the external working temperature and conditions in which the plant functions. It is perhaps worth mentioning that compressors utilised in refrigeration plant are generally expected to comply with the strict requirements of underwriters' codes, classification societies, and insurance companies, and J & E Hall compressors are built with this need very much in mind.

Compressors making up the Veebloc range form the most important group built by the company. These compressors are of the single- or double-throw crankshaft type with three, four, six or eight cylinders, and there are four basic sizes which carry the designations V54, V92, V127 and V178. The complete range covers bore and stroke sizes from 54 by 51 to 178 by 140 mm, and in the heading illustration can be seen a type V127 8-cylinder compressor which has an overall weight of 1600 kg and is designed to operate at a maximum speed of 1000 rev/min.

Veebloc compressors are of totally-enclosed construction and the entire interior of the crankcase and cylinder housing acts as a container for the refrigerant, which is most commonly either ammonia or Freon. The 'searching' nature of the refrigerant makes it essential that the compressor is leak-proof to a high degree and this feature is of particular importance with the crankcase, which must be sound and free from micro-porosity.

Preparation of crankcases for machining

Veebloc compressor crankcases are of grade 17 cast iron and are supplied to the company in an unmachined condition. All machining operations, with the exception of the preliminary milling of datum faces and drilling and reaming of datum holes, are carried out on NC machines and some idea of the amount of machining involved can be obtained from the finish-machined V178 crankcase shown in Fig 1.

Crankcase castings are machined on NC machines, for which it is required that workpieces are of reasonably consistent dimensions. To ensure that any dimensional variations are allowed for before machining is undertaken, purpose-built fixtures are used for checking and marking out castings

---

The use of NC facilities is appropriate to the production of such components which are required in continually-recurring batches and with which numerous faces, holes and bores need to be machined. In machining large and complex cast components under numerical control, however, problems can often arise as a result of dimensional variations with the as-cast parts.

Such dimensional variations render difficult the at-

1 *This finish-machined cylinder block and crankcase is the major component of a type V178 Veebloc compressor which, when assembled, weighs more than 3000 kg*

incorporates form plates corresponding with the shapes of the areas that require to be machined

2 *Large crankcase castings are dimensionally checked and marked out with the aid of this fixture, which*

1

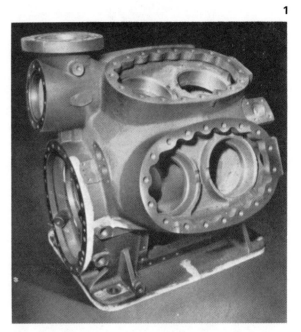

2

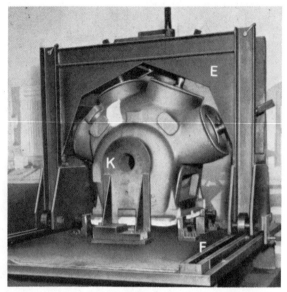

3

4

3 *The bridge structure on which form plates are carried moves on rails which enable it to be brought over the casting being checked*

4 *Small crankcase castings are checked and marked out using fixtures and templates of the type shown here*

5 *Kearney & Trecker Milwaukee-Matic Modu-Line machining centre, on which the smaller crankcases are machined*

5

tainment of the close dimensional control and repeatability which are essential features of NC. If the variations are of sufficient magnitude, it is often not possible for appropriate account to be taken unless there is intervention by the operator during the automatic cycle; consequently, unless the precise nature of the variation is known beforehand there will always be a danger of components being scrapped.

All crankcase castings are subjected to close inspection on being received and are then marked out in preparation for machining. To ensure that significant dimensional errors are precisely determined at this stage, all castings are checked and marked out

with the aid of purpose-built gauging fixtures, which are so designed that they enable all aspects of a casting to be examined visually at a single setting. As a result, the amounts of metal to be removed from different positions on a casting can be optimised, to ensure that machined surfaces will clean up fully. Use of the fixtures also provides the benefit of reducing the time taken for an operation which previously created serious restrictions in the progressing of castings to the machine shop.

Such checking and marking-out fixtures of two basic types are employed by J & E Hall, and the unit to suit crankcase castings of the larger sizes (for type

6 *This Burkhardt &
Weber MC4 machining
centre has a heavy-duty
boring and milling spindle
and a twin-spindle indexing
housing, and is equipped
with two independent tool
magazines. One of these
magazines serves the heavy-
duty spindle, and is shown
to the right and has
capacity for tools weighing
up to 80 kg*

6

V127 and V178 compressors) is seen in use in Fig 2. A casting to be checked is mounted on a table resembling a surface plate, and there are jacking screws as at *A* and pusher screws as at *B*, for alignment vertically and horizontally, respectively. The position of the casting longitudinally on the fixture is determined with reference to two setting members indicated by *C*. These members are of L-shape in plan, and the short legs (which extend radially outward relative to the castings, are required to be engaged by the inner faces of two arcuate lugs on the casting, as indicated by *D*. These setting members are removable from the fabricated column on which they are mounted. During all subsequent alignment movements of the casting prior to marking out, the positions of these faces relative to the fixture must be maintained.

The main checking assembly of the fixture takes the form of a bridge structure *E*, which is mounted on rails attached to the table to enable it to be brought into position over a casting and locked into place by means of pins as at *F*. The arrangement is seen more clearly in the rear view of the set-up in Fig 3. The bridge structure carries individual form-checking plates as indicated by *G, H* and *J*, which are used for aligning and marking out in connection with the various cored holes and their surrounding faces. These and other checking plates such as that indicated at *K*, Fig 3, are removable, and there is a certain degree of interchangeability to cater for both 6- and 8-cylinder forms of the casting within the one bridge structure. Similarly, interchangeable bridge structures permit two sizes of casting to be mounted on the one surface plate.

With the checking plates in the working positions, a casting can be set by means of the jacking and pusher screws to occupy the optimum position for machining all the surfaces, and in this position it is marked out. The last surfaces to be marked out are those of the base *L* and end face *M*, Fig 2, which when machined (on a conventional horizontal borer) form the datum surfaces in connection with all subsequent NC machining operations. Any areas of the casting which are found during marking out to deviate significantly from the dimensional tolerances required by the programme are marked accordingly; in this way the danger of a casting being scrapped during machining is considerably reduced.

Marking out of castings of the smaller sizes is carried out with the aid of fixtures of the type shown in Fig 4 on which a casting is mounted with its main axis vertical. The fixture has been designed so that the various cylinder arrangements available for a given size of compressor can be checked at a single set-up. Detachable form plate gauges of the type indicated at *N* are placed on machined pads on the fixture base, where they are positioned by means of radially-extending tenon slots.

Machining and tool changing

Machining of the base and end face datum surfaces on a casting is performed on a Jameson boring machine, which also provides for drilling two clearance holes and drilling and reaming two datum holes, in the base of the part. All subsequent machining operations are carried out under tape control on one of three machines, the choice of which depends largely on the size of the casting. Common fixturing permits a degree of interchangeability between machines. Two set-ups are required to complete the entire machining of each casting; one for all operations on the vertical faces and the other for those on the inclined faces.

The smaller castings, that is those for the V54 and certain of the V92 compressors, are machined on the American-built Kearney & Trecker Milwaukee-Matic Modu-Line machining centre shown in Fig 5.

One of three NC machines is employed for all but preliminary machining of the crankcase castings, choice depending basically on the size of the casting. Brief details are given of the tooling arrangements for these machines

---

This machine was installed in 1970 and has a single tape-controlled rotary table and a Bendix control system. A chain-type magazine provides capacity for 50 tools, which are selected on a random basis by means of the well-established Kearney & Trecker coded-ring identification system. Maximum tool weight as used by J & E Hall is 30 kg.

Medium-size castings, that is for compressors up to that designated V127 and in particular for the V92 range, are machined on the Burkhardt & Weber MC4 machining centre in Fig 6. This machine was installed in 1969 shortly after its introduction and provides for point-to-point working in conjunction with a tape-controlled rotary table, and a Siemens control system. A feature of this machine that has proved of great value to J & E Hall is its tool-holding capacity, which enables a wide range of tools to be accommodated including boring bars of very large size.

On the MC4 machining centre, the head incorporates a heavy-duty spindle for milling and boring, also an inclined indexing unit that houses two spindles for drilling and other operations. For machining, this housing is set with one spindle extending horizontally to apply the tool it carries to the workpiece, and the other spindle is then positioned vertically upwards. Automatic tool-changing is undertaken in connection with the latter spindle, while machining is in progress, and indexing of the housing through 180° is then all that is required to bring the fresh tool into use. Tools are stored in a chain-type magazine, with capacity for 36 units, that is between the twin columns carrying the head. Automatic tool changing is also obtained with the heavy-duty spindle, which is seen in use in the illustration, and tools are stored in a circular indexing magazine which has a capacity for 15 tools that can be of very large size and weigh up to 80 kg. Arranged horizontally, this magazine is seen at the right in Fig 6, and a number of the large tools that are shown mounted in it have bodies of aluminium. Further reference to this practice will be made later in the article.

8

The largest crankcase castings produced are those for the V178 range of compressors, and are machined on a Giddings & Lewis-Fraser type 70A horizontal boring and milling machine having a 5 in spindle (Fig 7). On this machine, in contrast to those discussed above, all tool changing is performed manually, with the aid of a small jib crane where necessary. As can be judged by the casting shown mounted on the rotary table in the illustration, some of these tools are of considerable size. Again, a rotary table is provided on the machine and with the set-up here shown, it carries a fixture with which the base is formed by upper and lower plates whereon the opposing faces are inclined at corresponding angles. The upper plate can be swivelled on the lower member. With this arrangement, the fixture can be set as seen, for operations on vertical faces on a workpiece, and turning the upper plate through an angle of 180° provides for those on a face that is inclined at an angle double that at which the plate faces are inclined.

Tooling

Reference has already been made to the size of a number of the tools used in the machining of the crankcases, and an indication is provided by Fig 8, which shows a tool trolley loaded with some of the larger units that are required. The boring bar indicated at P is of the type employed for operations in very large bores with diameters up to 500 mm, and it is of non-circular design in order to minimise the weight. Also in the illustration can be seen boring bars of two other types: in the lower positions at R and S are associated roughing and finishing units, each of which will machine a 4-step bore in one operation. The roughing tool S is shown again in Fig 9 mounted in the Microfix unit that is used for presetting all the NC tooling. The largest multibore tool

---

Very large tools are used for certain machining operations and to ensure minimum weight use is made of Duralumin for the main bodies

---

T (Fig 8) is designed to machine in one pass, a bore with three steps up to 250 mm diameter; this tool is equipped with Valenite E-Z Set cutter assemblies.

It will be readily appreciated that if constructed entirely of steel such tools would be very heavy – and in certain cases the weight could be beyond the load capability of the machining centre tool changers. The problem is particularly acute in connection with the G & L-F horizontal borer, for whereas tools are changed manually, the machine is used for the very largest castings, requiring tools which – if of steel – could weigh in the region of 100 kg. For these reasons, the practice has been adopted extensively of constructing the bodies of tools from Duralumin aluminium alloy, which for a given volume has a weight that is less than that of steel by a factor of 3. Since the use of aluminium alloy in this way is somewhat unusual, the decision was taken only after extensive trials relating to tool stability and deflection, also to ascertain the possibility of problems resulting from heat transfer or the bedding of the steel components into the softer body of the aluminium alloy when under load. Such trials indicated the aluminium alloy to be a suitable

7 *The largest castings are machined on this Gidding & Lewis-Fraser horizontal boring and milling machine, with which tool changing is carried out manually*

9 *For presetting tools used on all three NC machines at J & E Hall works, this Swiss-made Microfix unit is used*

8 *Because of the large size of many of the tools required in machining compressor crankcases, aluminium alloy is used extensively for tool bodies*

9

material and the practice has now been applied for five years. A further advantage gained with aluminium-body tools is the comparative ease with which the bodies can be machined, thus facilitating tool construction, and reducing lead times.

Tools required for the NC machining operations on a specific casting are selected by the programmer, who compiles a list of the items needed as part of the programming operation. This list is then provided on a drawing that forms instructions for the NC machine operator, and which also indicates the workpiece and its relationship with the machine datum. Tools for which drawings already exist can readily be selected and prepared by the tool setter from drawings contained in the tool stores. Where a tool of a fresh design is required, an appropriate drawing is prepared by the jig and tool drawing office and issued for production. A tool assembly drawing is sent to the tool-setting area, and it includes information concerning the length and/or diameter to which the unit must be set. This drawing also lists the components that are required for the assembly, and specifies the type of tipped tool to be employed. The latter information is provided largely for the purpose of rationalising and controlling the stocking and use of tip tools. For storage, large tools are not always dismantled to the same extent as smaller sizes, but the cutters are generally mounted in exchangeable cartridges which permit quick changeover. Normally, three sets of consummable tools are prepared for a given batch of workpieces, to enable those forming one set to be in the course of being sharpened while a second set is cutting and a third set is being held in reserve. After use, all tools are inspected, sharpened, and dismantled for return to the store. ⊙

A die casting machine is here seen supported on four Hillman rollers equipped with swivel plates

NC Master Family Programming

By R. W. Nichols
NC-GT Co.

Master Family Programming (MFP) is Group Technology (GT) applied to N-C computer programming. It is especially suited to management's need for multidiscipline control to reduce costs and manpower and increase production through standardization of improved machining methods, tooling and equipment.

MFP reduces writing many single part N-C programs, eliminates tape and tool proofing as well as programmer documentation.

In this presentation, MFP methods are detailed and illustrated for use by N-C programmers.

MFP will achieve savings with increased productivity and unusual synergistic benefits.

WHAT IS MFP

MFP IS the creative manager's answer to mass production using N-C equipment. MFP uses GT geometric composites along with coordination of other disciplines. Altogether they provide a superior use of N-C computer programming techniques, as well as fully using the computer print-out capability. This is required to eliminate programmer documentation. The MFP (software) is a permanent program made up of many modules or single elements. These modules are:

1. Heading, part number, post processor call-out, and plotting.
2. Variable input
3. Geometry
4. Tool data (cutting tool off-sets)
5. Constants, also fixture off-sets, machine origin
6. Feeds/speeds tables
7. Macros two up to five
8. Written set-up instructions and/or pictures of set-up and tooling carbide selectors.
9. Machine motion sections (cutter path) with operator and inspection instructions and cutting tool call-outs. The use of logic and logic statements which link these various sections together.

HISTORY OF MFP

MFP originated about 1960 and has grown slowly until the recent popularity of group technology (GT). MFP development has been mainly through the efforts of the G. E. Corporate Technical Services and a few others. MFP has been kept under wraps, so to speak, because it gives you unfair advantages over competition such as:

1. one day tape and parts delivery;
2. expansion of product sizes and material types without tooling or programming delays;
3. and, justification for new N-C equipment.

Some unusual applications of MFP have been the elimination of the N-C programmer. Another case is where the design engineer and N-C programmer developed an MFP where a new code number is added for each new configuration, doing away with six thousand drawings. This code is the input to the MFP for manufacturing the part via tape. In 1969 there were over three thousand tapes made from one packing ring family and not one bad tape.

There are two pseudo types of MFP currently being used.

1. The most popular conception of MFP, or something like it, is the use of a series of fantastic macros.
2. A few clever programmers have made some improvements to reduce the programming task by adding feed/speed tables, operator instructions to the tape listing, coded tools on listings and some macros.

The REAL MFP takes these modules and a lot more and cuts the N-C programming and manufacturing costs to the bone. We will describe the futuristic MFP with CAD/CAM a bit later.

The MFP disciplines are:

1. Engineering Design
2. Manufacturing Engineering
3. Planning
4. Tool Design
5. Programming
6. Tool Presetting
7. N-C Shop Operation

With this in mind it is especially important that management understands, supports and monitors the MFP efforts, just as you would for any new N-C equipment or other large capital investment.

MFP is a discipline of GT -- GT (Group Technology) IS the manufacturing concept in which similar parts are grouped together to be produced more efficiently. Specifically, GT is the coding and classification of parts that are related, similar, look-alikes, or have common traits. This coding makes possible the most economical handling in the design and manufacturing cycles, helps assign work to the most suitable machine, and even calculates in advance what the machining time will be. GT establishes cost savings in nearly every area in your company.

GT takes time, money, training, and according to many experts in GT, it requires about eighteen months before you get to the N-C equipment with the tooling and set-up benefits it gives.

N-C machines are high priced and certainly have a much higher ration of productivity than conventional equipment and typically get a lot of attention by management, as they should. With the high dollar investment it makes a lot of sense to start using GT disciplines via the MFP in one to three months instead of eighteen months or more. MFP would load your N-C machines with proven tapes and methods. The cost savings would start increasing sharply when compared with the normal GT approach.

We have not seen MFP at a facility where GT is in use. Also, of the many companies we have talked with, only once were we asked if GT was needed. Our answer was we don't need GT, but it would simplify working with MFP.

MFP is especially suited for use in the manufacture of the following products:

1. Own products (proprietary)	7. Bearings
2. Oil Tools	8. Printing Machines
3. Gears	9. Off-the-road heavy duty equipment
4. Valves	
5. Pumps	10. Turbines and turbine engines
6. Shafts	11. Tooling

MFP is a MANAGEMENT TOOL - WHY - because MFP is a multidiscipline system. When implemented it will maintain better control than any cost reducing system that we know.

The reason MFP gives you this control is when the program is completed a programmer can't make tapes if any one discipline has changed. Of course, if there is a necessary change such as tooling, added processes, engineering, etc., it costs money. Only now manufacturing engineering or the design engineering departments, who were responsible for the change, will have to give you the hours (money) for programming, and now you, the managers, are made aware of the impact on your schedule and budget.

Another example of MFP benefits would be -- you need a new VTL N-C machine to handle increased production. Using MFP we never give a second thought to planning, tooling or programming support functions because we have one or more MFP for the VTL's that can produce tapes to machine every standard valve body condiguration engineering can produce. In fact, we could buy two or three more VTL's and never have a problem supporting them.

At CCI we average approximately one or two parts per tape/listing. One programmer, with the help of a terminal operator, can produce 15 to 40 tapes/listings in one day using MFP besides his regular work. We call these achievements synergistic benefits.

MFP is sometimes difficult to understand. You have heard about standardization of designs, cost savings, manpower reduction, management control, but to implement, it takes a "true believer' with a sincere approach and follow-up. If this type leadership is not available for MFP, your N-C programmers will have high blood pressure and down the tubes goes MFP, your future for GT CAD/CAM, and possibly the whole integrated computerized manufacturing system concept. In other words, MFP is a tool for managers to achieve multidiscipline control.

MFP justification by management can be just one or all of the following:

1. Increase N-C programmer productivity, or keep from adding manpower, or in exceptional cases replacing the programmer with a lesser skilled person.
2. Improve N-C machine utilization with less hazard to the machine and operator, and reduce tape and tool proofing.
3. Reduce computer costs
4. Increase productivity by applying GT to the N-C programming.
5. Improve quality of product
6. Increase management control of product design, tooling, and manufacturing methods using closed loop systems.
7. Reduce in-process inventory
8. Reduce scrap

STANDARDS OF MFP

The implementation of MFP with its multidiscipline integration creates synergistic benefits through the standardization of:

1. Methods
2. Equipment
3. Cutters
4. Fixtures
5. Materials
6. Planning
7. Programming
8. Feeds/Speeds/Stepovers
9. Time Standards
10. Product Design

Do you now have such standards? If you do have standards, can you make them work? The chances are you may not have all of these standards, nor control of them. MFP establishes these standards and maintains them.

COST SAVINGS AND BENEFITS WITH MFP

Management

1. Increases control of costs, methods, tooling, manufacturing and capital equipment purchases.
2. Standardizes product design
3. Reduces in-process inventory
4. Reduces scrap
5. Modernizes and updates for computer integrated manufacturing systems (CAD/CAM).
6. Eliminates collision on N-C machines due to programming errors.

Manufacturing/Industrial Engineering

1. Standardizes manufacturing methods/sequences/words.
2.   "   equipment types and sizes.
3.   "   holding fixtures and chuck jaws.
4.   "   cutting tools reducing inventory.
5. Reduces paper/documentation and storage area.
6. Produces accurate machining times.
7. Provides automatic machine justification.

Manufacturing/Shop

1. Eliminates time (normally 20-40%) used for tape and tool proofing.
2. Permits rapid loading of N-C machines with proven tapes.
3. Eliminates risk of machine collision due to tape error.

N-C Programming

1. Simplifies a complex N-C programming task into a clerical function, which can be completed in minutes.
2. Eliminates documentation made by programmer for shop (pre-set, set-up and operator instruction).
3. Reduces single part programming time by using MFP proven modules, i.e., feeds/speeds, macros, origin statements, cutter off-sets, operator set-up information, etc.
4. Reduces tape change time for engineering changes.
5. Capability to provide same day new production tape and listing.
6. Reduces computer costs.
7. Eliminates plotting for production tapes.
8. Standardizes feeds/speeds/depths of cuts (stepovers) for each material and for each size machine.

9. Eliminates programmer time used for tool and tape proofing, which normally is 20-40% of N-C machine time.

10. Increased capacity to handle peak programming loads.
11. Reduces programming manpower.
12. Provides the necessary time for programmer to optimize moves and feeds/speeds.
13. Upgrades programmer's skills for application of group technology to other areas.

The N-C programming seldom has any new programs run through the computer which produce a useable tape the first time. A typical program of several hundred lines has some typing, key punch or punctuation errors. After these minor errors are corrected, we may have a few major type errors that either stop the program from running or may be in the wrong coordinates. MFP eliminates all of these errors.

We frequently find, in single part programming, three days or more are needed to debug a new tape for a safe run. This doesn't include fine tuning of approaches, clearances where you cut air, feed rate fixes, and sometimes 10, 20 or 30 parts are completed while still debugging. MFP bypasses the normal program preparation errors and machine time for debug of feeds/speeds, depth of cuts, RPM and SFM. This will produce a major impact on the N-C operation.

The most unusual advantage in the MFP justification is no tape or tool proofing -- can you believe it -- usually a month or two is required to establish shop confidence using MFP tapes. Also, no first part scrap; run one part per tape for spares or production.

I would like to briefly mention MFP with CAD/CAM, in which you can produce pictures, as well as tapes/listings/inspection/tooling and planning copies in a single computer pass which takes seconds, or a minute or two if you are using a stand alone mini-computer. This GT CAD/CAM was accomplished three years ago and is one way to achieve a totally integrated computerized manufacturing system. We mention this to increase your awareness of the power that the MFP concept can achieve in a parametric CAD application. This type of software program would help reduce industrial "future shock".

The start for MFP is made by anyone with knowledge of the product and the N-C programmer collecting all available blueprints. Then an evaluation of these drawings is made to:

1. select families-of-parts, group by configuration;
2. select N-C machine type and size;
3. group into material types;
4. draw composite of all configurations that make up the family-of-parts;
5. list all variable inputs;
6. decide machining sequences;

7. select cutters to be used;
8. select tooling and design, or purchase as required.

Selecting family-of-parts (item #1 above) is unnecessary if you have GT installed at your company.

MFP is basically an N-C computer program for families-of-parts, so you can use any computer language that has logic and jump-to statements and computer listings which use title or print statements, as used in APT N-C programming language.

The APT, ADAPT, UNIAPT languages we know work. They are taught in the public high schools and colleges, and are supported by industry and government. There are thousands of post processors written in APT for your machines and there are support personnel in many locations throughout the U.S. and the world.

#1 HEADING

The first step in MFP is the heading which has the part number, letter change (configuration control), operation sequence number, equipment name or code number, part name and tape number; this is a single line and yet this is where you start standards.

The next line is the post processor call-out.

On a new MFP you may need to add a plot statement on the next line. Plotting is needed for complex geometry, where there are many logic statements which jump past various program sections. When we first started we had several MFP's that ran perfectly for nearly a year, and finally we used a new section and there were several corrections that could have been made initially if we had used a plotter. The programmer should input variable that cause each section (cutter path) to be used and plotted. The reason for this is there can be two or even twenty different configuration, some may not be used for months, and the plotting will debug the cutter path for all configurations.

#2 VARIABLE INPUT

The input section is the only part that requires coordination with the manufacturing engineer because of material types and sizes furnished, the raw stock diameters, pre-cut lengths, flame cut diameter, forged billets and premachined part dimensions. The manufacturing engineer and planners sometimes do not understand the seriousness of wrong material size and types on N-C machines, especially lathes and VTL's. We have had this problem. It was finally solved by:

1. The MFP listing prints out the diameter and the cut-off length (including multiple parts, i.e., where one bar or slug makes two or more parts), so now the foreman and operator can check for proper material size the tape is made for and not call the programmer.
2. Inform management about N-C concepts and that the tapes cannot accept excessive material sizes without N-C machine crashes.

A fixed description is needed after each input which describes what the variable inputs are. On most inputs a sketch/picture is also added to give clarity. This picture should include the maximum/minimum diameter, length and each material type for which the program is made. This makes it possible for machine tool planners to make inputs. Remember, you may not use the MFP for months, and we don't rely on memory.

One item of special attention is that every input should be a drawing dimension. We have found that programmers, being very smart people, sometime use short cuts on inputs. It takes awhile to stop this practice. It is much quicker to check inputs that are blueprint dimensions. Using this concept you may eliminate the programmer (think about that).

The computer is made to compute so we try never to manually calculate any dimension. This is important to carry through because you may have two or twenty programmers working on many programs, and they all should be using the same approach to maintain clarity and simplicity, which is the hallmark of MFP.

#3 GEOMETRY

The third program section is very familiar to programmers -- it is geometry. It looks the same as single part programs but with several changes. First, it is documented in true computer programming procedure for reference to the print, and each group of symbols should be uniformly sequenced by number. The only differences will be the use of logic (which is rarely, if ever, used in the geometry section), and the use of variable input symbols in place of dimensions.

You start by taking a xerox copy of the geometry composite and identify each line, radius, curve, taper, etc. It is a very organized systematic approach. Each horizontal line is called HL1, HL2, etc., starting at top, each vertical line starting from the center line is VL1, then the chamfer and taper line TL1, then the radius R1, etc. However, we were to talk about logic.

With the use of logic in the geometry section, MFP becomes MFP with group technology. Without logic, an MFP will machine any part that is in that family of one configuration, regardless of size or material.

However, with logic in the MFP you can run not only any size part in the family, but you can also change to various different configurations, some of which were not in your original family-of-parts. All of a sudden we have found an additional benefit in MFP.

#4 CUTTING TOOL DATA

The fourth section is the cutting tool dimensions and description data. In most companies you already have a pictorial coded tool book. The N-C machining centers, almost without exception, need one because you may have several hundred cutting tools for one N-C machine. It gives you a retrieval system through coding and grouping. Most of us have been using this GT discipline for years. It is the serious approach to reducing cutter documentation and an absolute must for MFP.

Here is a sample for VTL, boring bar, turning tool and drill.

```
$$ *****       TOOL LIST VTL       *****
T105XA = 3.281      $$ O.D. TURN NEG. HOLDER, 15 LEAD CLW
T105YA = 15.704     $$ INSERT 90 DEGREES
T105RA = .046
T211X = 0.0         $$ 5" DIA. SPADE DRILL, CLW
T211Y = 26.975      $$ TURRET POS #1 OR #5
T211R = 0.0
T301X = -3.02       $$ I.D. BORE, CLW, TURRET POS #2
T301Y = 20.343      $$ INSERT 60 DEGREE POSITIVE HOLDER
T301R = .032
```

You need the double dollar sign after each line of data with a description, also spindle direction and insert code and carbide grade for debugging and easy reference to your tool book. The X and Y values should be added to the coded cutter book, and then it can be used for single part programming information. You may want to include all the tools to be used on each N-C machine, so this tool data module can be used in any program. The calculation of cutter data off-sets and coordinates are now eliminated, and this module can be used for all programs to follow, including conventional single part programs.

When you complete this tool data and debug your programs, you now know what your standard cutting tools are. Now is the time to start weeding out, reducing cutter inventory and applying GT techniques. This data in most computer applications for N-C programming can be permanently stored in a system macro and eliminated from the MFP. Then a single line call statement will bring all the data to the computer when the program is run. This can be applied to F/S tables, constants, macros and set-up instructions with pictures.

#5 CONSTANTS

Constants are the smallest module in the MFP, but of key importance. The constants are the inputs to the spindle speed change macro needed for start-up, and are set low for built in safety. Two constants are needed for approach clearances. The clearance dimensions vary from .06 to .200 or more depending on the size of the machine and material variations.

The two key items in this section are:

1. Home positions for lathes for Z axis and X axis are required for both incremental and absolute programming.
2. When absolute programming is used as illustrated, it will produce Z zero at the end of the part when programming for chuckers, lathes of VTL's. Then all the moves which cut, drill, bore, etc., into the part are Z minus values and are the same as the dimensions shown on the blueprint. This also applies to the X axis. Now, fitting the tape output to the blueprint is one important step to gaining shop confidence, helping to train new N-C operators, and making it much easier to debug programs.

A sample for lathe.

```
$$ ***** CONSTANTS *****
SP1 = 200              $$ SPINDLE SPEED SFM OR RPM
DIR1 = 0               $$ CW SPINDLE
VLL = 2                $$ RADIUS FOR TLCHG MACRO
DRIL = 0               $$ 0 = SFM, 1 = RPM
PR = .125              $$ APPROACH TO CUTS
PR1 = .250             $$ APPROACH TO CUTS CAST PARTS
OR = .062              $$ OVERRUN ON TURNING
ZAXPOS = 50            $$ FLOATING Z AXIS HOME POSITION
T1P1 = POINT/(ZAXPOS-(D17 + CONS3)),-15.250,0   $$ HOME
CLR = LINE/PARLEL,VL1,XLARGE,PR
RG1 = 1                $$ SPINDLE RANGE
MAX1 = 290             $$ MAX. RPM IN SPINDLE RANGE
```

#6 FEEDS/SPEEDS TABLES

The next section is the feeds/speeds table and is one area that the manager and tool engineer should have total awareness of because this is the most important money making module in any MFP.

The cutting/feeds/speeds, stepover tables for each material and each size machine for operations such as single point threading, tapping, rough and finish boring, grooving, drilling, counter sinking, rough and finish turning for materials such as inconel, titanium, 4130 CM, 1020 CR, aluminum brass, ductile iron, stainless steel and many others. There is one table for each material the program makes parts for, also

we add one or two extra tables for much faster debug of feed and speeds.
Each MFP should have three or more of these tables.

Feed rate override is rarely needed in MFP except upward over 100%.
The feed/speed tables may take time, but when you have it, it is like money
in the bank. You have just stopped all tool engineers, programmers and
foremen from changing feeds and speeds and from now on they no longer
need to be concerned about chip load, feed and speed, H.P., etc.

To simplify, we will use three tables -- one for 316 stainless,
1020 CR, and one for 4140. Let's pick a VTL to show more range in the
use of feeds and stepovers. The first is the feed rate and we find that
four is usually enough: F1 = .030 roughest feed, F2 = .022 semirough,
F3 = .009 finish, F4 = .005 fine finish. Now, if we were on a chucker
we would need F5 = .003 extra fine, and in a rare case we could have up
to seven separate feeds. Now, think about it, do you really need more
than four feeds?

Next is RPM, usually six to eight and then SFPM, which typically
needs only four.

And now the depth of cuts, called stepovers, four is also a standard;
however, five or six are sometimes needed. You can start with two
extreme stepovers. The first is the heaviest and that is usually .400 up
to 1.125 on a VTL. The smallest depth cut should not be less than the
radius on the tool, say .045 or .065 on a VTL. Now we can select two
more in between, say .100 and .250. This table is made for ASTM A105
(1030) low carbon steel, and now we make two more tables, one for ASTM
A182-F11 (4041) C.M. and one for 316 stainless steel. Now we have three
tables which is usual for most MFP, but we could have six or seven
tables to also include inconel, copper and brass, etc., etc.

```
        $$ ***** FEED AND SPEED TABLE *****
            IF(MAT)IDD1,ID19,ID20
    ID19)   F1=.022        $$ ROUGH FEED: FOR STEEL A105 (1030 CR)
            F2=.014        $$ ROUGH OR FINISH FEED
            F3=.010        $$ FINISH FEED
            F4=.007        $$ FINE FINISH FEED
            SF1=750        $$ 380 RPM X 6' DIA. 230 RPM X 10' DIA
            SF2=450        $$ 280 RPM X 6' DIA. 175 RPM X 10 DIA
            SF3=260        $$ 160 RPM X 6' DIA. 100 RPM X 10 DIA
            SF4=230        $$ 150 RPM X 6' DIA. 90 RPM X 10 DIA
            SF5=200        $$ 130 RPM X 6' DIA. 75 RPM X 10 DIA
            SF6=160        $$ 110 RPM X 6' DIA. 60 RPM X 10 DIA
            SF7=130        $$  85 RPM X 6' DIA. 50 RPM X 10 DIA
            SR1=40         $$ RPM X 6" DIA=35 SFM, 10' DIA 50 SFM
            SR2=70         $$ RPM X 6' DIA=95 SFM, 10' DIA 150 SFM
            SR3=120        $$ RPM X 6' DIA=190 SFM, 10' DIA 280 SFM
            SR4=180        $$ RPM X 6' DIA=280 SFM, 10' DIA 480 SFM
```

-continued-

```
        DC1=.400      $$ DEPTH OF CUT
        DC2=.300
        DC3=.250
        DC4=.050
        JUMPTO/ID21
ID20)   F1=.018       $$ ROUGH FEED: FOR F-11 MAT'L (4041 FORGED)*
```

We are illustrating a VTL or large lathe type, but with machining centers you may have to enter all the RPM's in the table to get the correct RPM. The minimum use of three material tables will be very useful when debugging. An example, everything may be too fast on the RPM or SFM and the cuts are too deep, so you input the next slower table, or someone will come up with a new material for the family-of-parts program. The chances are one of the tables may do the trick or you may need to add one new table. The new table would take only thirty minutes to an hour or so, and then you could make a tape very quickly.

When setting up an MFP for the threading you should include five or more feeds/speeds tables. Some day the tool room or R&D will need a shaft to be threaded of aluminum or inconel, but the MFP may have only feeds/ speeds tables for copper and 1020 C.R. steel. These extra tables can really make the programming department out a hero because you can output a tape on the spot, like in fifteen to thirty minutes. On thread programs we include OD and ID. Also, you can include an alternate use of a tap on small to medium internal threads in place of single pointing. Usually the post processor will need to be upgraded for tapping -- they rarely do tapping.

These feed and speed tables are advantageous during debug. As an example, you probably will need to change one, and it could affect hundreds or even thousands of feeds on one tape because you use only four feeds, four RPM, four SFM and four depths of cuts. This allows you very rapid turn around during debug because all you do is change let's say F2=.012 to F2=.010 and now every place F2 is used the feed output is changed.

#7 MACROS

Let's talk about macros, but first let me say macros are only one small section, and in a few cases we never use macros.

The one word best suited to explain MFP is logic. Even a short program would have twenty to fifty logic statements, with big programs two or three hundred uses of logic may be required. The other reason we don't use macros is especially important. Macros cost more, much more to run through the computer. With logic you jump through the program at the lowest possible cost. We have one large MFP (4000 lines) that outputs fifteen to twenty blocks (4' to 8") of tape for a large expensive valve body at four to five dollars per run. This gives you an idea of how cheap it is. However, this same 4000 line MFP does output tape over

2000 blocks (150' to 400') at seven to thirteen dollars per run. This would not be possible using macros.

We do use macros for repetitive items. Starting an MFP we don't always know what is to be repetitive. However, for sure we do have two repetitive items -- changing of tools and spindle speeds. There are other items used less, but used in a macro.

These are the basic macros used primarily for lathes, chuckers and VTL's.

1. Tool change, to send tool safely to home, change tool and return to the work piece.
2. Spindle RPM or SFM change, rotation direction.
3. Error macros to output an explanation to the programmer how the input exceeded the program, cutters, tooling or machine capabilities.
4. Positions programmable tailstock
5. Bump stop positions material on chucker for single parts.
6. Bar puller positions material on chucker.
7. Threading single point

A review of macros used in an MFP for a lathe reveals that only numbers one and two for tool change and spindle change would always be used and then maybe one or more of the remainder as required.

The macros we use are a little more detailed than average, but most programmers have written large macros. So these macros will not be difficult to write. One point about tool change macros -- there are no exceptions in the use of cutter tool book code numbers. This is part of the bare bones requirement to eliminate preset documents for each tape.

There is one other essential in the real MFP, and that is the system capability to produce a computer listing with APT print and title type output. As far as we know, only one or possibly two systems do not have this capability, but there is no doubt they all could if they understood your requirements. An interesting fact is that computer service companies make more money on the main frame than on the terminals supporting edit mode, input and printing listing. So MFP helps you and them to use your time and their equipment at the lowest costs.

#8 N-C OPERATOR SET-UP

The goal of the MFP is to let the computer do the work regardless of what it is. This includes set-up, preset, operator and inspector instructions. This instruction varies with each family-of-parts and type of N-C machine. On the VTL we may have a tool selector which prints out on the listing each tool, as coded in the tool book, starting with the longest bar, then the turning tools and their position on the turret. We usually include a carbide grade selector that typically uses C2, C5 or ceramics, but it could have more if needed. This is for the pre-setter and machine operator, and is common to pre-set documents.

One of the nicer things we have the computer do is output a picture of the set-up. This absolutely clarifies the set-up for the operators and foremen who may have trouble reading long set-up instructions. The computer picture can include dimensions, part shapes, tool names such as chuck, brand name, model number, jaws, tool number, grid position for start-up, axis direction, machine home, etc.

This section is where we add the material size if it is needed. This usually applies when there is no control over material size, such as forged billets and oversize bar stock. We print out the maximum diameter and length the tape will make. These print-outs are output from the input data and are no extra work by the programmer. Now you have relieved the programmer of being asked: Will this tape make a part with this odd size material? This also helps management fix the problem (be it system or people).

#9 PROGRAMMING WITH LOGIC

Now that we have covered all the elements, let's start programming the machine motions or cutter path. We'll start with a 10" diameter bar stock (or it could be forged billet) by 12" length, and we will do a series of cuts using the logic method. The sample part variable inputs are:

```
ROD = 10        $$ ROUGH OUTSIDE DIAMETER
 OD =  8        $$ FINISHED OUTSIDE DIAMETER
LGT = 12        $$ FINISHED LENGTH OF PART (PER B/P)
```

This is how they would appear in an MFP.

The start is a series of input to the tool change and spindle change macros to select:

1. RPM or SFPM
2. Spindle speed (RPM or SFPM)
3. Direction of spindle
4. Spindle range (1, 2, or 3, etc.)
5. Maximum RPM constraint (for G96 function only).
6. The input variable to position the cutter from the centerline of the spindle.

Sample program using logic:

```
SP1 = SF4
DIRI = 0
VLL = ((ROD * .5) + PR)
RG1 = 2
MAX1 = 776
CALL/ASTER
PPRINT T100 POS #2 OFFSET #2
PPRINT ROUGH TURN O.D.
CALL/ASTER
CALL/TLCHG,POS=2,TX=T200X,TY=T100Y,$
HCODE=2,RC=T100R,LOC=TO,CHG=0
DEP = 0
I = ((ROD-OD)*.5)

ID50)   IF(I-(DEP + DC4))ID52,ID52,ID51
ID51)   DEP=(DEP + DC4)
RAPID
GO/(L1=LINE/PARLEL,HL2,YLARGE,DEP)
FEDRAT/F3,IPR
TLLFT,TOLFT/L1,(LINE/PARLEL,VL1,XSMALL,((LGT * .5) + .25))
GODLTA/0.,.06,0
RAPID
GO/PAST,CLR,PS
JUMPTO/ID50
```

The important presetting cutter documentation are the cutter tool code number, the number of the turret and the off-set number. The next line is a description of the machining process, i.e., "rough turn excess stock".

The call statement for the tool change macro is next. You insert the cutter tool data (see section #4) for the X, Y dimensions and the radius. The off-set number is inserted and then the position of the cutter -- to, past or on the radial location from the centerline of the spindle. Then the macro is instructed to send the tool home first and index, or index in its current position.

We are ready to use the MFP method to cut metal. Let's assume we have already faced off the end and now are ready to remove 2" from the OD. First, we establish a counter. We call it "DEP" and give it a value of zero. Now in the case of our sample, we have a 10" stock diameter and an 8" finished diameter, so we have 2" which we divide in half to equal 1". We call this value "I".

Now we use a logic statement which asks three questions: If "I" is less than, the same as, or more than our counter "DEP". Now if "I" is less than or the same as "DEP" we jump over the rough cutting section to the finish section which we turn in two cuts. The first pass leaves .06 to .09 and is made at a heavy feed rate. The finish cut feed rate is selected to produce the proper finish. But "I" has a value of 1" and "DEP has a value of zero ("I" is more than "DEP") so we drop to the next line.

Now we select a stepover or depth of cut code from our feeds/speeds tables (see section #6). Let's decide on DC1 (DC1=.300). We add the depth of cut value (.300) to "DEP" which now makes "DEP" equal to .300. The first move is a rapid .300 past the stock OD. Previous to this the tool change macro has positioned the cutter in front of the work piece by .06 so the cutter does not hit the material. The actual cutting is the next move. Let's use F1 (F1=.018) feed rate (1PR). At the end of this cut we make a .020 to .040 move away from the metal and then rapid to a clearance in front of the part.

Now we jump back to the original logic statement and ask the same logic questions again. But now our counter "DEP" equals .300, so we repeat the same sequence over again, only .300 deeper. The computer repeats this sequence until "I" is less than or equal to "DEP".

The length of the cut is generally to a variable input, such as to a shoulder, taper or groove. However, it could be half the length (LGT=12.0 variable input) because the whole bar has to be turned to the finished diameter. So the length of the cut would appear as LGT divided by 2 + .125. After one end is finished we reverse the part in the chuck and repeat the operation on the opposite end.

This logic method can be applied in many other configurations, including some interesting deep drilling, and will result in reduced computer costs when compared to macros. This logic approach is much simpler to create and easier to change than macros.

Our experience is when you start using logic, there is seldom a need for the more difficult macros.

We have detailed the various elements required for MFP. There could be one or even two elements you may not need. However, there may be additional upgrades that become more customized depending on your product, N-C equipment, and computer support systems, including a system capable of interfacing to CAD when you use the MFP approach. Now you will really feel the synergistic impact of MFP.

Presented at Prolamat '79

Advanced Manufacturing Technology, P. Blake, ed.
North-Holland Publishing Company
© *IFIP, 1980*

CUBIC—A Highly Automated System for Programming NC Machining Centres

By H. Stoltenkamp, W. J. Oudolf
and H. J. J. Kals
Twente University of Technology
(The Netherlands)

This paper describes a software system for programming NC machining centres, which is in development at the Laboratory for Production Engineering of the Twente University of Technology. The system is limited to 2½ D control and includes the following operations: milling, drilling, boring, reaming and tapping. A high level of automation is reached by automatic tool selection and automatic generation of both tool path and machining conditions. Part geometry is described by means of shape elements which are composed into shape compositions. Measures are taken directly from the drawing; relation to the origin of the part coordinate system is done by the program. As a result of this, part coding is easy, no computer programming experience is required.
The system design is modular and well structured to further portability and flexibility. The program will be running in a time sharing environment.

1. Introduction.

CUBIC, a software system for programming NC machining centres with 2½ D control, is still in development at the Laboratory for Production Engineering of the Twente University of Technology. The system can handle flat, box-like and block-like parts and perform the following operations on them: milling, drilling, boring, reaming and tapping. It is a further development of a preliminary designed system which followed the basic philosophy of MITURN as designed by Koloc. Consequently it has the same advantages as MITURN has for the programming of lathes: i.e. easy part coding and a high level of automation in the field of technology.

Redesigning the original system was found necessary in order to meet demands for flexibility and computer independency. Additionally a capability for milling operations is being added.

Today the demands for both flexibility and a high level of automation are in most cases contradictory.

However, this contradiction is not a fundamental one. Rather this is caused by an as yet insufficient knowledge of software technology. But there are economic constraints too. In the design of CUBIC it is tried to meet both demands in an economic way.

2. Workpiece description.

The code in which the workpiece is described is largely numerical and looks rather incomprehensible at first sight. It is, however, easy to learn by workplanners even if they have no knowledge of computers or computer programming. Described are the shapes to be machined, which appear on the workpiece together with the positions on which they must be manufactured. The code contains no information about the sequence in which they will be manufactured or the tools that they will be manufactured with. This, together with the machining conditions, is generated by the system.

A complete workpiece description consists of four code blockes: i.e.

- The Header code block containing general data regarding to the workpiece, like number of drawing, workpiece material, postprocessor to be used, etc.
- The Shapes code block describing the shapes to be machined
- The Coordinate code block in which the necessary coordinates are coded
- The Application code block in which the relation between shapes and the position on which they must be machined, is laid down.

```
10/  # H
20/    TEST, A CODING EXAMPLE
30/    1075156, STEEL 60
40/    KTMM, POSTPROCESSORNAME
50/    .05,10.,1.,MM                          * QUALITY
60/    1,23                                   * CLEARANCE PLANE
70/  # F                                      * SHAPE COMPOSITION
80/    1,1,1,10,0,0,0,25,1,0,0                * HOLES UPPER LEFT
90/    2,1,1,8,0,0,0,15,1,0,0                 * HOLES UPPER RIGHT
100/   3,1,11,15,0,0,0,10,8,4,1,2,3,4         **
110/   3,2,12,15,0,0,0,-10,1,1,2              *
120/   3,3,12,15,0,0,0,-10,1,2,3             * SLOT WITH
130/   3,4,12,15,0,0,0,-10,1,3,4             * COLLISION SIDE
140/   3,5,12,15,0,0,0,-10,1,4,1             *
150/   3,6,19,5.5,21,5                        **
160/   4,1,5,8,1,0,24,0,2,0,0                * THREADED HOLES M8 WITH
170/   4,2,9,19,32                           * COLLISION CYLINDER
180/   5,1,11,16,0,0,0,5,9,5,4,6,7,8          **
190/   5,2,12,16,0,0,0                        *
200/   5,3,12,16,0,0,0,-5,5,6,7              * POCKET
210/   5,4,12,16,0,0,0,-5,5,7,8              *
220/   5,5,12,16,0,0,0,-5,5,8,5               **
230/   6,1,11,0,0,0,0,1,10,4,10,11,12,13 * PLANE ON RIDGE
240/  # A                                     * MACHINING POSITIONS
250/   1,1001,LIN,14,15,4,20                 * 4 HOLES UPPER LEFT
260/   2,1002,ARC,16,3,0,90                  * ARC UPPER RIGHT
270/   3,1003,PAT,1                          * SLOT
280/   4,2004,GRD,17,18,19,3,2,25,25         * 6 THREADED HOLES
290/   5,2005,PAT,5                          * POCKET LEFT
```

```
300/    6,3006,PAT 10              * PLANE
310/    1001,4007,TRT,14,160,-180,-24,90,4    * HOLES ON RIGHT SIDE
320/    2004,2008,TRT,17,135,-50,0,90     * THREADED HOLES RIGHT
330/    2005,2009,MIR,20,1          * POCKET RIGHT
340/ # C                   * COORDINATES
350/    1,12.5,21,6.5,11.34,21,1
360/    2,0,1,15,1,0,1,1
370/    3,95,2,0,2,0,1,1
380/    4,0,3,0,1,0,1,1
390/    5,-45,20,10,21,19,21,1
400/    6.0,5,69,62,21,0,5,1
410/    7,46.19,30,6,1
420/    8,0,7,0,5,0,5,1
430/    9,0,5,0,5,-5,5,1
440/    10,0,21,140,21,55,21,1
450/    11,0,10,30,10,0,10,1
460/    12,180,10,0,11,0,10,1
470/    13,0,12,0,10,0,10,1
480/    14,20,21,210,21,0,1,1
490/    15,20,14,0,14,0,14,1
500/    16,140,21,180,21,0,1,1
510/    17,10,21,130,21,0,5,1
520/    18,25,17,0,17,0,17,1
530/    19,0,17,-25,17,0,17,1
540/    20,90,21,0,21,0,5,1
550/    21,0,0,0,0,0,0,1
560/    22,0,1,0,1,-10,1,1
570/    23,0,21,0,21,60,21,1
580/ # E
```

Fig. 1   Example of workpiece coding

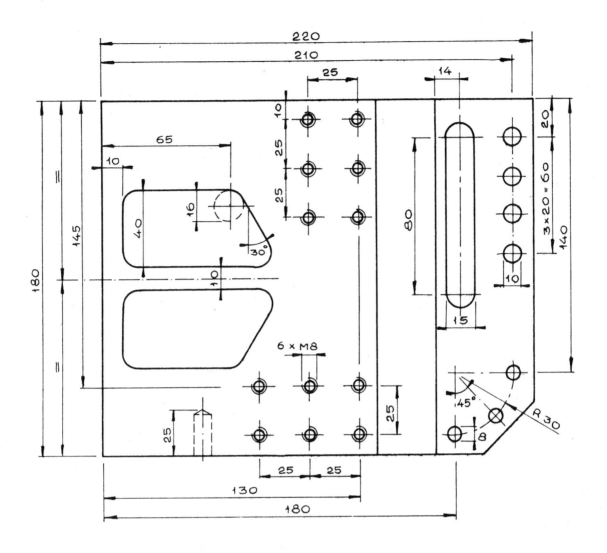

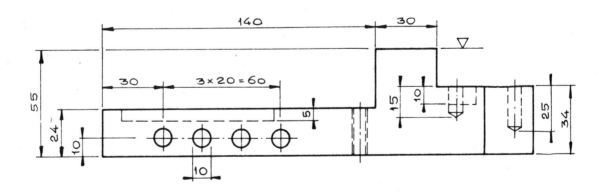

Fig. 2   Workpiece coded in fig. 1

The code blocks are seperated by delimiter records and may appear in any order in the input file.

Delimiter records are those records that have a cross-hatch (#) on the first significant position, they are:

\# H to denote the Header code block

\# F to denote the Shapes code block

\# A to denote the Application code block

\# C to denote the Coordinate code block

\# E to denote the end of the input file

The Shapes to be machined are coded in the Shapes code block as shape compositions which are built from shape elements of one shape element group. These conceptions require proper definition:

Shape element:

Every part of a (composed) shape defined as such.

Shape composition:

Every defined composition of shape elements belonging to one   shape element group.

Shape element group:

A specified group of shape elements.

There are two different shape element groups i.e. hole elements and flat elements. Examples of the former are the round cylinder and the threaded cylinder. Examples of the latter are the flat and the side.
Figs. 3 and 4 show these shape elements and some shape composition which can be built of them. In the Shapes code block each shape element is coded in one record. A shape composition consists of a number of records with the same shape composition number (first number of a record).

The relating of the measures on the drawing to a coordinate system usually requires a great deal of calculating work by the

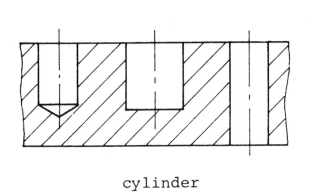

cylinder

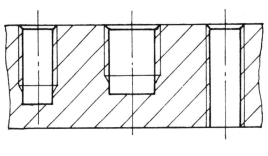

threaded cylinder

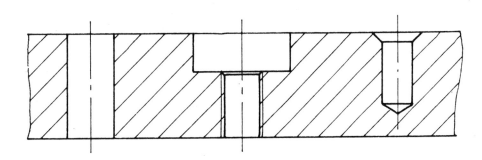

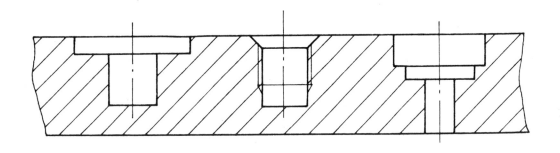

Hole elements and some possible
shape composition

fig. 3

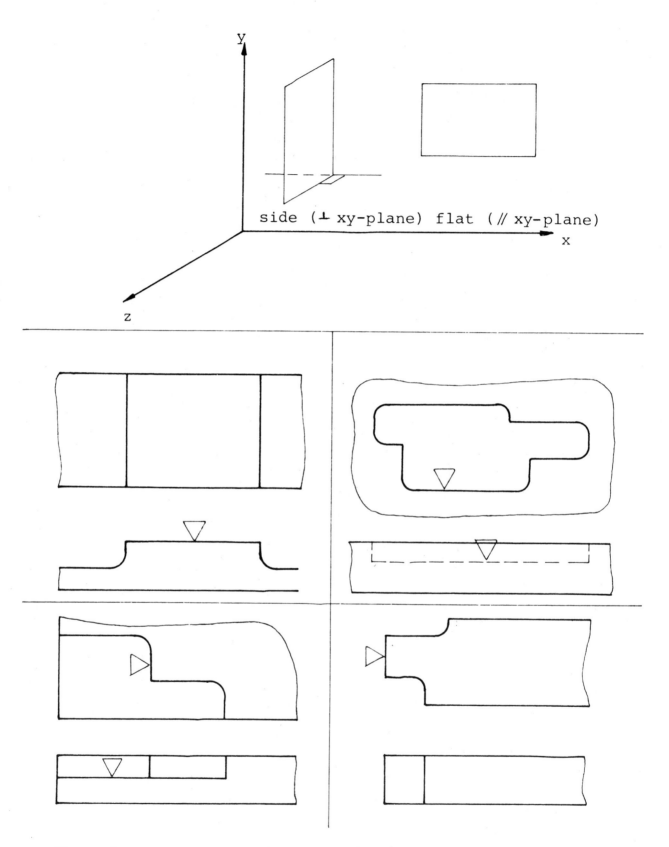

side ($\perp$ xy-plane) flat ($/\!/$ xy-plane)

Flat elements and some possible shape compositions

Fig. 4

part-programmer. This is due to the fact that the designer deals with functional measures, like the distance between two cog-wheel shafts, rather than with measures related to any coordinate system. However the relating of measures to a given coordinate system can easily be automated, thus allowing the part-programmer to enter the coordinates in the Coordinate code block in relation to one another as indicated by the drawing. In this way arithmatical work in part-coding is minimized if not totally annihilated. Fig. 5 shows how this is done. As an example let us take the following chain of x-coordinates. Point nr. 4 has an x distance of 55 from point nr. 1, which is -30 from 2, which is 0 from 3.

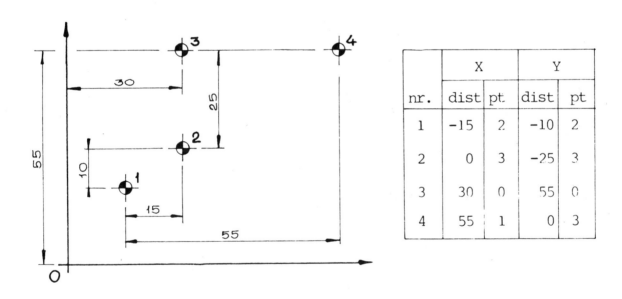

nr.	X		Y	
	dist	pt	dist	pt
1	-15	2	-10	2
2	0	3	-25	3
3	30	0	55	0
4	55	1	0	3

Fig. 5   The coding of coordinates

Point nr. 3 refers to the origin of the coordinate system and therefore has absolute coordinates. The end of such a chain is the point that refers to the origin. If, due to a coding error, there is no such point then the chain is not complete and absolute coordinates cannot be calculated. This does result in an error message.

In the Application code block the shape compositions are arranged into patterns with the help of the coordinates from the Coordinate code block, thus placing them in the right position on the workpiece. All shape compositions must appear in the Application code block. If not, they will not be manufactured. The arranging is defined by means of four pattern statements each denoted with a three character keyword.

In addition two copy statements, equally denoted by keywords, are available to copy predefined patterns.

The keywords for the patterns are:

LIN to denote linear patterns

GRD to denote grid patterns

ARC to denote arc shaped patterns

PAT to denote randomly arranged patterns

The keywords for copy statements are:

TRT to denote translation and rotation

MIR to denote mirroring

The meaning of the Application code block lies in the fact that in using pattern- and copy-statements the Coordinate code block becomes much shorter. For example, a grid pattern of say a hundred holes can be coded in three coordinate records and one grid pattern record, once the shape composition is defined in the Shapes code block.

3. Part processing.

The manufacturing procedure to make a given shape composition is determined, taking into account the nature and the dimensions of the shape composition as well as the required accuracy and surface quality. A manufacturing procedure is defined as a collection of machining operations required to manufacture a given shape composition thereby obtaining the required quality - e.g. the manufacturing procedure to manufacture a threaded hole consists of the machining operations centering, drilling and tapping.

The tools are selected with machining operation and dimension as keys. The procedure allows equal tools being selected for different shape composition (e.g. a straight hole of 8 mm. diameter and a threaded hole of M8 may require the same drill). In such cases all machining operations requiring equal tools are arranged under one physical tool in order to reduce presetting time and tool-changing frequency. The minimization of the number of tools requires a predefined loading sequence of the tools in the spindle of the NC-machine, to ensure a proper sequence of machining operations on one specific shape composition (e.g. the system should not try to tap thread in a hole that has not been drilled yet). Therefore tools have been assigned a type number. The tools will be sorted in an ascending order with type number as primary key and diameter as secondary key.

In general the proper sequence of machining operations when dealing with flat, box-like and block-like workpieces is:

1. face milling
2. pocket and contour milling
3. centering
4. drilling
5. boring
6. countersinking
7. reaming
8. tapping

Tool type numbers derived from this sequence of operations can be:

1. end mills
2. slotting end mills
3. shell end mills
4. centre drills
5. drills
6. core drills
7. boring tools
8. counter bores

9. countersinks
10. reamers
11. taps

This tool-sequence is not unique. In some other cases one might need a different sequence. For instance a slotting end mill can also be used for drilling.
One solution is that one physical type of tool will be assigned more than one type number. The machining operation and/or the shape composition will determine which number has to be used.

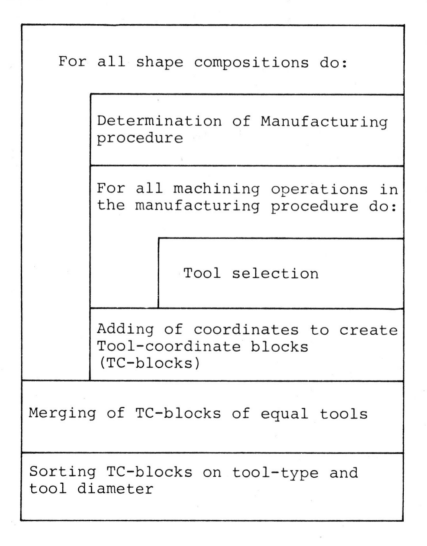

Fig. 6   Nassi-Scheiderman diagram depicting tool
          selection and sorting.

The rules for the selection and ordering of tools are laid down in relational tables which are stored in files outside the program text. These files are called Manufacturing Procedure files. They can easily be changed by means of off-line utility programs. With this feature the workshop experience about the system performance can be fed back into the system, thereby improving it. This is one of the ways in which the contradictory demands for flexibility and a high level of automation can be met.

The technology contents of the system comprising tool path generation in milling and the adjustment of the different machining values in terms of feed, speed etc. is very much dependent on the distinct machining operation. In the case of threading or finishing operations like reaming, the applied method for the selection of the values of the machining variables is rather straightforward. Only fixed values are fed into the program.

In the case of milling and drilling, however, the system design includes modules for process optimization on economic principles. The applied method corresponds to the one used in a formerly designed program system for turning which is described in ref. (3).

4. The collision problem.

The tool path over the workpiece is generated by the system and is therefore unknown to the workplanner. This, together with the fact that only those parts of the workpiece which have to be machined are described in the input, causes the possibility of collision of the tool with obstacles on the workpiece, like ridges, clamping devices etc.

A fully automated solution of this problem is only possible if the complete shape of the workpiece is described. Since this would lead to a considerable extension of the inputfile we have settled for a lesser level of automation in this respect. In the solution presented here the workplanner has

to recognize potential collision dangers and take certain
measures in order to avoid them.

The collision problem can be divided into two parts:
- collision danger while moving the tool in z-direction or
  while the tool is cutting;
- collision danger while positioning in x and/or y-direction
  in rapid traverse.

In the first case the collision danger is caused by tool-
clamping devices which may collide with an obstacle close
to the shape composition. This problem has been solved by
introducing two dummy shape elements, which define the free
space around the shape composition.
The two dummy elements are the hole element collision
cylinder and the flat element collision side. Fig. 7 gives
an example of both.

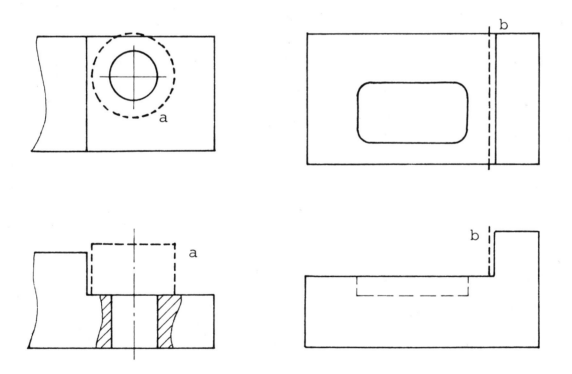

Fig. 7 - Dummy shape elements to prevent collision
         a - collision cylinder
         b - collision side

During positioning in x and/or y direction the collision problem consists of the tool colliding with an obstacle on the workpiece.

To solve this problem, the workplanner has to define area's on his workpiece within which there are no obstacles. These area's are called pattern groups. If the tool travels out of one pattern group into another, the system will assume that an obstacle exists in between and the tool will be retracted to a clearance plane which is defined in header code block. To avoid unnecessary retraction, pattern groups can be defined overlapping each other.

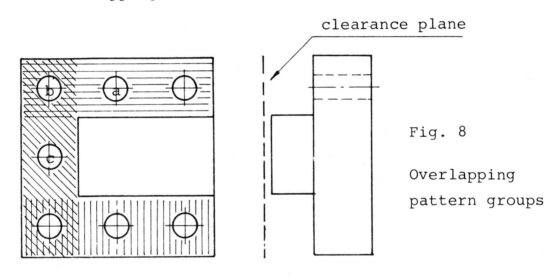

clearance plane

Fig. 8

Overlapping pattern groups

If, in fig. 8, a tool travels from a to b it will remain in the same pattern group and will not be retracted. During the same movement the tool enters into another pattern group, so if subsequently it travels from b to c it is not retracted either. If the tool travels directly from a to c it is retracted to the clearance plane because it is entering a different pattern group while leaving the other.

5. System design. (fig. 9).

CUBIC consists of a number of independent programs called modules. These modules will automatically be loaded and executed once the system has been started.

Advantages of this type of design as compared with linkage

in overlay mode are:

- Small units. Simple and cheap maintenance. Possibility of implementation on small computers.
- A simple break-restart possibility between modules.
- The possibility for (limited) user interference between modules.

A background memory was designed in order to save internal memory space and to pass variable data from one module to another. In this background memory all data are represented in two dimensional matrices of variable size called data-blocks, and stored in the background memory-file. A module can communicate with the background memory by means of sub-routine calls. These subroutine calls perform such functions as reading, writing, creating data blocks etc.
Housekeeping within the background memoryfile is done auto-matically .
Another feature of the system design is represented by the fixed files in which all workpiece independent data are stored. The workpiece independent data include the data about machines, tools, materials and manufacturing procedu-res. By storing these data outside the actual program text it is possible for each of the users to have his own set of fixed files. Changing of the contents of these files is done using off-line utility programs. In this way the program system becomes better accessible for general use and more flexible with respect to an individual users' needs (see paragraph 3).

The Tool-Coordinate file or TC-file represents the post-pro-cessor interface with the system. Another utility program is available to inspect the file and change it if the results are not satisfactorily.

The communication of all modules with the different files is controlled by means of subroutine calls as is the case with the background memory file. Every module of CUBIC therefore consists of two parts as is shown in fig. 10.

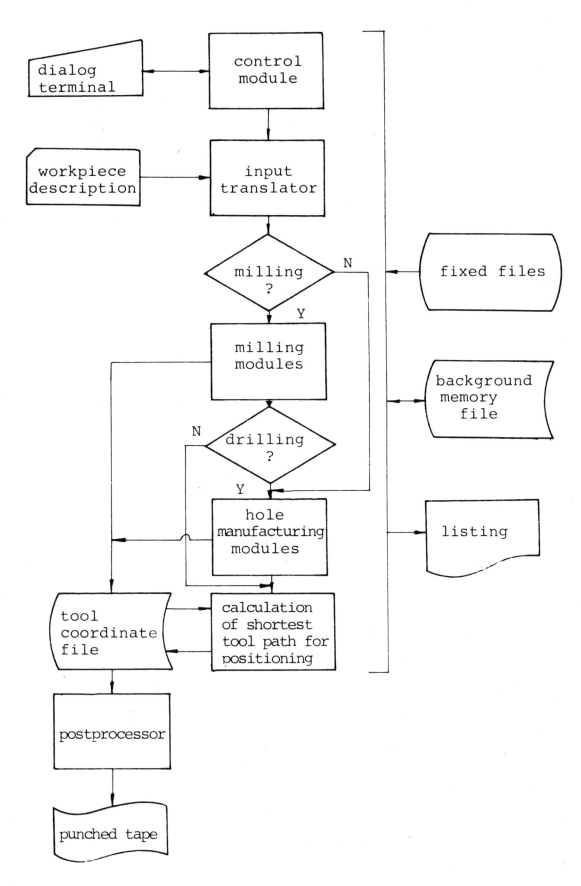

Fig. 9   Schematic system layout of CUBIC

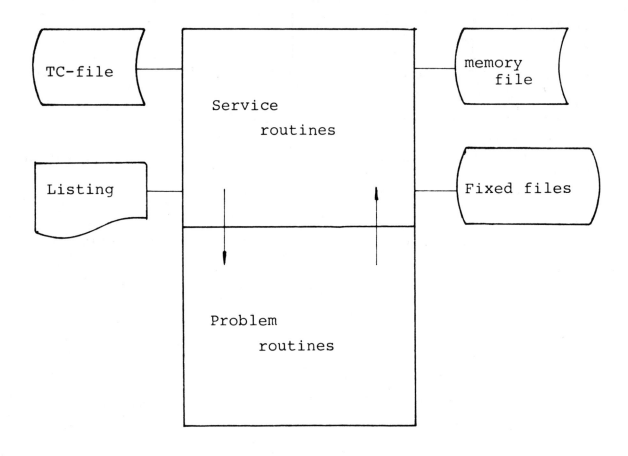

Fig. 10 Schematic layout of a module

The service part contains the subroutines which perform the communication between modules and files and between the different modules.

The problem part contains the routines that handle the problem for which the module was created.

6. The position of   NC-program systems like CUBIC in the production environment.

Program systems like the one dealt with in this paper are very complex and time consuming in development. Standards of flexibility and accessability have to be set to guanrantee a general use because the potential individual users are not in a position to develop there own programs. Experience has learned that the need for automatic part programming is particularly high in average to small size workshops, dealing with average to small lot sizes. In these cases hand programming becomes quite too expensive. Also intolerable losses by drop outs, when the production of a new lot is started, can frequently be observed. Multiple use of a program system by many workshops introduces the problem of the use of many different tools on many different machine tools in many different environments. On the other hand the complexity of the machining processes and the difficulty in describing them has put us in a position in which we have to put up with very simplified process modelling and are unable to deal with the influence of the dynamic behaviour of machine tools on tool life and that of tool materials on surface roughness, to mention only two problems. This means that fundamentally and practically none of the available data are generally applicable and that data cannot be exchanged when some condition is different. In this respect the creating of large data bases can be considered as a prove of our incapability rather than as a solution to the problem. Today the only way out seems that every user has to create his own data, based on the experience in his own workshop.

Indeed it is quite evident that the use of NC-program systems in cases where process technology is important, like in turning and milling, can only be succesfull if the problem of the determination and collection of reliable data has been solved.

A second problem arises where the servicing of NC-program systems is concerned. They need continuous maintenance for two reasons. First there is the fact that even large field tests are never prove for a proper working for the most different machining conditions existing in different workshops. Secondly, these systems need frequently to be adapted to new developments in course of time. Both facts imply that a seperate system management by experts is required.

A third problem concerns the low standardization level applied in numerical control. This is reflected in the large variety in the existing postprocessors. For this reason post-processors necessarily are part of the manufacturing system. Fig. 11 shows the schematical design of a production information system - i.e. the manufacture information part only - in which allowance is made for the above. Advantage is made of the modular setup of the NC-program systems, which means that the different files are part of the data base of the production system, while the program library is separated and may be used by many factories in time sharing. On the other hand the post-processors are stored in the data base.

The figure shows three feed back modes
1. the feed back of production flow information (instantaneous)
2. the feed back of actual time- and product quality data (time standard and tolerance evaluation with e.g. cumulative averaging).
3. a heuristic feed back (process technological data).

The last mentioned feed back loop contains product quality data and tool utilization and thus process utilization data.

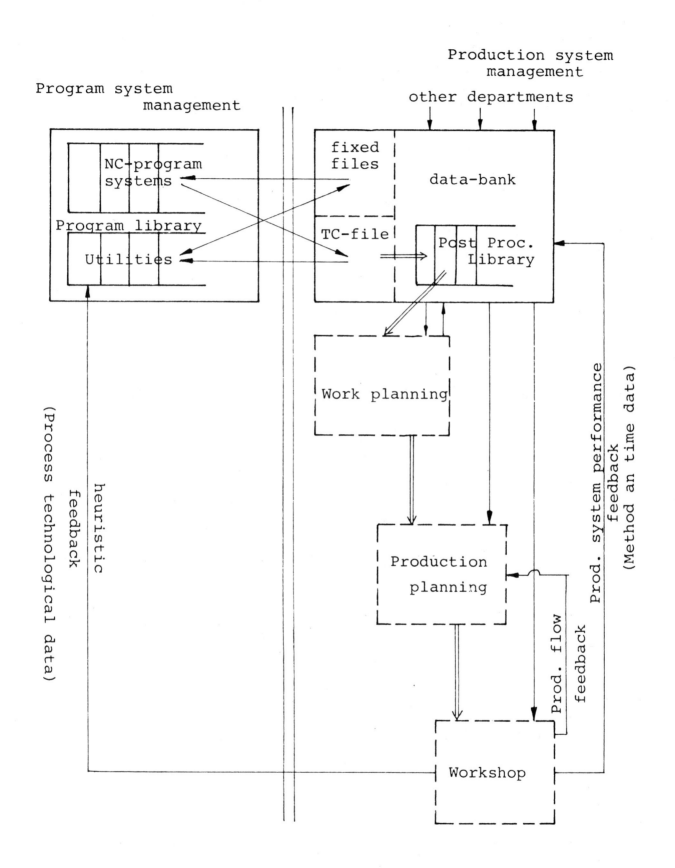

Fig. 11 A production information system

These data are being compared with the previously calculated values in the TC-file stored in the data base. The process technological data in the fixed files are subsequently changed in order to minimize the difference between calculated and actual results. All this is done by a utility program which is part of the program library.

So, starting from a rather limited data file of the program system, this is the way in which every user is able to get his data adapted to the conditions in his own workshop.

References:

1. W.J. Oudolf, Development of a Programming System for NC Machining Centres. Annals of the CIRP Vol. 25-1-1976

2. J. Koloc, Practical aspects of computer-aided programming of NC lathes. Proceedings of PROLAMAT 1976

3. H.J.J. Kals, J.A.W. Hijink, A Computer Aid in the Optimization of Turning Conditions in Multi-Cut Operations. Annals of the CIRP Vol. 27-1-1978

Presented at Prolamat '79

Advanced Manufacturing Technology, P. Blake, ed.
North-Holland Publishing Company
© *IFIP, 1980*

Application of PEARL For Programming 3-D Measuring Machines

By T. Pfeifer
and D. Loebnitz

Technical University Aachen (Germany)

Numerically controlled measuring machines are increasingly used for measurement in production processes. They can only be used economically, if the appropriate software is provided. In this respect two software problems have to be solved:

first, the programs to control the machine and to evaluate the measured data
second, the task of programming the measuring procedures themselves. In this paper, a development is discussed which allows the measuring procedures to be formulated in a high level language similar to EXAPT. The basic software for this programming system on the machine part is written in PEARL, which allows parallel processing of different tasks and thus increases the overall speed of the system.

1. The Problem

The demand for increased economy and flexibility in process-measurement particularly is small-scale and medium batch production has led to the development of three-dimensional measuring machines. In this manner it is possible to measure proportions which either cannot be measured at all or only with elaborate efforts.

Although revolutionary in the field of metrology, measuring machines are not quite a novelty. As far as their structure is concerned, they are conparable with machine tools, e.g., milling or grinding machines.

Like machine tools, measuring machines have a coordinate-system represented by three orthogonal axes. With each point of the working area of the machine, a unique triple of values x,y,z is associated, by which the position of the tool can be represented. On measuring machines, the tool is replaced by a probe, which gives a signal on contact with the workpiece. There are different degrees of automation of measuring machines, ranging from manual inspection machines to motor-driven, computer-numerically controlled- (CNC-) measuring machines. The development described below refers only to CNC-machines.

Besides the apparent parallelism between NC-machining and NC-measuring there are several differences. One main difference is that the geometry of the workpiece is generated by machining, while it has to be inspected during measuring. In this procedure several measuring points on the surface of the workpiece are registered and the values describing the geometry will be derived from the measured points by computations. In this way, a more or less complete model of the workpiece can be built within the computer. This model is generated by using statements uncommon to machining procedures.

As in the case of machining, the problem concerning the programming of measuring procedures is also to be solved here.

While standardized commands and high-level-programming languages for formulating programs in NC-machining exist - e. g. APT- like- languages, EXAPT - measuring procedures have to be programmed according to the teach-in method. In principle this means, that the operator controls the machine for a sample workpiece and communicates instructions to the computer via pushbuttons. The computer stores the intermediate as well as measuring positions and the actual problem into its memory and can then execute the "taught-in" solution for all the following workpieces. The disadvantage of this programming method is that during the learning phase the machine is busy with being programmed, and not with measuring, as it should be. This leads to rather expensive programming costs. Besides, a sample workpiece is necessary, and this cannot always be provided during initial phases of small-scale production.

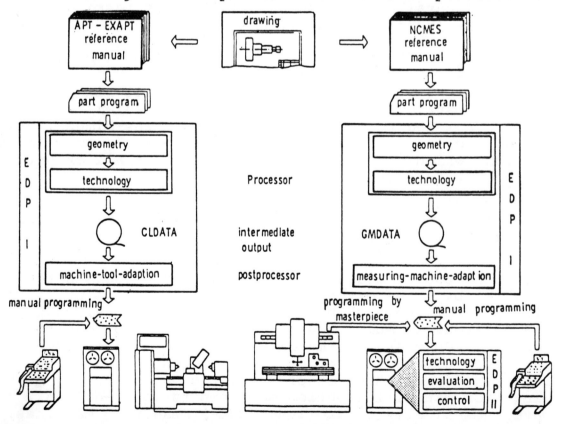

FIGURE 1

The time required to generate a ready measuring program lasts until the sample workpiece is actually produced. This period may be quite long, so that it is possible that several waste parts are produced before the measurement results are available. To overcome this problem, a command structure has been defined, which permits measuring programs to be punched on tapes, that can be later read and executed by the computer at the measuring machine. Because of the analogies to NC-machining, the command format of the punched tape was chosen according to DIN 66025. In sprite of this, all the evaluation commands regarding measuring had to be redefined. The next step was the development of a high-level programming language. Thus it is no longer necessary to program a measuring procedure step by step. With the aid of a few complex statements it is possible to develop complex measuring programs, e.g. the distribution of measuring-points in a hole to measure diameter and position of the hole. Such a language has been developed in the Federal Republic of Germany, sponsored by the government, it is named "NCMES":"Numerical Controlled Measuring and Evaluation System" and is now being tested in the industry.

2. The System NCMES

The language NCMES is a high-level problem- oriented language for NC-measurement of workpieces on three dimensional measuring machines. It is closely related to the APT- like language EXAPT for machining, the translation procedures are also similar:

- The programmer writes a part program according to an inspection plan and part drawing of a certain workpiece.

- The part program is processed by the NCMES-Compiler, which is written in FORTRAN and available on CD- and IBM-computers. The compiler checks for formal errors and for collision between probe and workpiece, generates additional intermediate positions as well as measuring movements and generates the output on a punched tape, which is closely related to DIN 66025 (resp. the according ISO-standard) with some extensions particular to measurement.

- The punched tape is used as an input for a desk-calculator or process-computer at the measuring machine. Depending on the nature of the letters used in the commands, the latter are interpreted either as control-instructions for the machine or as calls for special evaluations programs.

The two main features of this concept are:

- the translation of the part-program to the punched tape; this is done off-line, i.e. without using the machine.

- the processing of the punched tape by the machine (and its controlling computer).

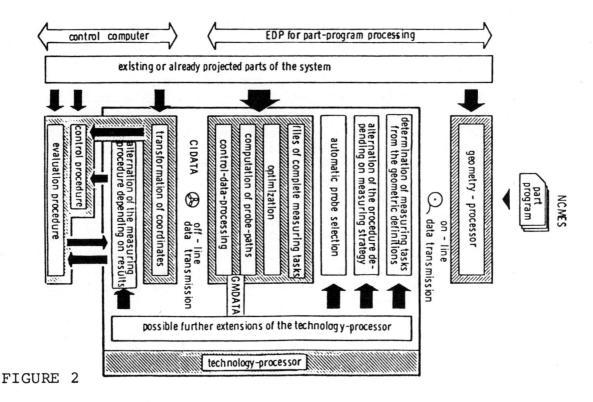

FIGURE 2

The communication between the two systems is done via punched
tapes. The following explanations are restricted to the
second feature, since the first has been described earlier /1/.

3. The Language NCMES

Like any other programming language, NCMES has several
different types of statements:

Geometric definitions, by which the ideal geometry of the
part is communicated to the computer. The computer requires
this to generate commands for traverse paths and to check
for collision. Geometric definitions are only needed for the
translation.

Technological statements, that give information as to which
probe is to used, what dimensions the probe have, and how
the workpiece has to be aligned. This information is partly
passed on to the measuring machine.

Commands for the machine´s traverse paths, by which the
machine is explicitly positioned, e.g. for reaching special
intermediate positions. Positioning commands are contained
in a part program to some extent and are passed on to the
machine, but most of the positioning commands on the punched
tape are generated by the translator from complex measuring
statements.

Evaluation statements, by which the computer is told what to
do with the measured points or the computed geometric
elements. These commands have to be passed to the measuring

machine because they cannot be processed.

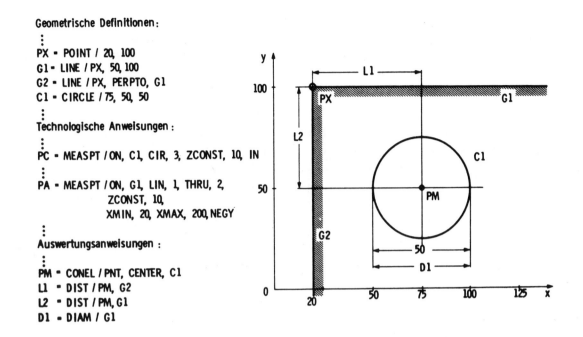

FIGURE 3

The language NCMES contains further statements that are a combination of several statements which are often used together. Such statements reduce the total number of statements to be written.

The current version of the system contains following geometric elements:

- point
- line
- plane
- circle
- sphere
- cylindre
- cone

Further elements are to be added.

These geometric elements can be combined not only to compute relations, distances and angles, but also for forming new elements, e.g. a line can be defined as the intersection of the planes.

4. Problems at the Machine

The NCMES-Compiler decomposes the statements of the part program into basic commands, generates positioning commands and evaluates all the geometric definitions. The output of the compiler is a punched tape containing all informations necessary for registering and evaluating measured values.

The program on the punched tape consists of

  -positioning commands
  -commands for calibration the probe (s)
  -commands for alignment, i.e. for establishing the relationship between the machine´s coordinate system and the one for
   which the program has been written.
  -evaluation commands

The positioning commands tell the machine where to go, - either
to intermediate or to meaduring positions.
The calibration of the probe is normally conducted by a
spherical gauge. The sphere is measured, and difference between
the actual diameter and the measured one issues the probe-
diameter. For determing the probe-data in case of several
probes, the relative position of the probes to each other is of
interest. Positioning commands are generated by computer and
given to the machine. After the probe data has been determinded,
it is stored on the disk for repeated use. In all measurements
the probe data is automatically compensated. The part programs
reference a coordinate system attached to the workpiece. The
commands to the machine have to be given in the coordinate-
system of the machine. Since the coordinate-system do not
coincide, the commands coming from the punched tape have to be
transformed before they can be sent to the machine. The
determination of the transformation is done by manual
measurements on the workpiece. Certain defined elements - holes
or planes - are measured manually and their position with
reference to the programmed coordinate system is known. The
transformation can now be determined - this is called soft-
warealignment.
The main task of the computer at the machine consists of the
evaluation of the measured data. In this respect the following
operations have to be made:

  -Comprising operations, which reduce the measured coordinates
   to the characteristic values of geometric elements, e.g. for
   a hole diameter and centre.

  -Combining operations, which join geometric elements to new
   ones, e.g. intersecting line between two planes.

  -Computing operations, which compute the relationship
   between geometric elements, e.g. the distance between two
   holes.

  -Comparison operations, which compare the actual values with
   nominal values and tolerances.

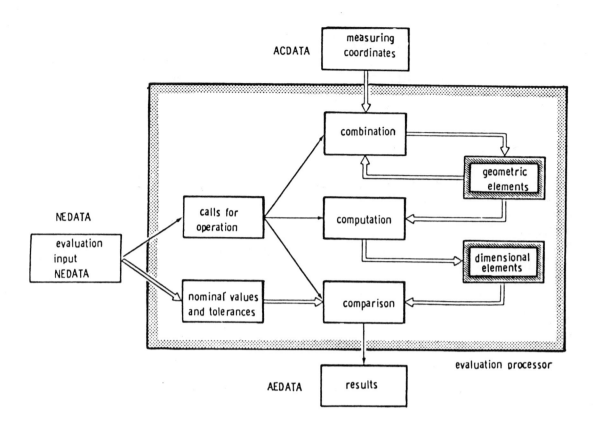

FIGURE 4

5. Measuring a Workpiece

In measuring a workpiece, the following operations have to be made:

- The probe data is determind, i.e. the probe (s) is/are calibrated, or, if the probe-data is already stored, they are fetched from the disk.
- The actual position of the workpiece within the working area of the machine is determined by taking measured values at predefined points.
- Now the measurement itself can begin; it consists of positioning and evaluating commands in random order; results are printed as soon as they are obtained.

Therefore the tasks of the computer at the machine are:

- Read and decode the punched tape; transform the positions.
- Give instructions to the measuring machine; wait for the answer from the machine and receive the measured values.
- Evaluate the measured values.
- Print a record.

These tasks have been realized in ANSI- FORTRAN as subprograms, i.e. functions and subroutines. Each task is started and executed sequentially to the end. The measuring time as a whole consists of the sum of

> time to read the punched tape
> \+ time to position the machine
> \+ time for computing the results
> \+ time to print the records

Since the computer is idle most of time - i.e. during the time the machine is running and during the input- /output-operations - the measuring time as a whole can be drastically reduced by parallel processing.

That means, that the measuring machine as the slowest part of all has to be kept busy. While the machine is running and the computer is waiting for the reaction, the next command can be read in, the results can be somputed and printed.

In this ideal case, the measuring time would be the sum of

> time to read the first command on the
> punched tape
> \+ time to position the machine
> \+ computing time for the last evaluation
> \+ printing time for the last result

Such a parallel processing cannot be achieved by ANSI- FORTRAN. Besides that, several problems arise, which are briefly discussed in the next chapter.

6. Problems of Parallel Processing

If several programs are to run simultaneously, it must be assured that they do not interfere with each other, viz. that they will have to wait for an event set by another program. In our system, since the measuring and the evaluating task work in parallel, care must be taken that the evaluation does not start before all the required values have been measured. On the other hand, the buffer for measured values must not be refilled before it is entirely processed. Beside this synchronization, the program must have the possibility of mutual influence, i.e. they must be able to activate, terminate, stop and continue each other. This is characteristic for the so-called real-time programming languages , which are being developed - e.g. CORAL 66 (Great.Britain), RTL (France), PEARL (F.R.Germay), US-DOD-development.

To program the NCMES-System, PEARL wa chosen because of its availability and non-dependency on a special computer.

7. PEARL

PEARL is a real-time language,i.e. the means to controll a process are integrated in the language definition. The main features are:

-Scheduling: the possibility to synchronize the computer´s program with the external process.

-Tasking: the possibility of programming parallel activities, which are independant of each other, but may influence each other

-Process- Input/Output: statements to transmit signals to the outside world and receive information from it by process-chanels

-Synchronization: the possibility to delay a task until an event caused by another task has occured, e.g. the filling of buffer.

-A clear distinction between the hardware- dependent and the hardware- independent part. The former is called the "system division", the latter "problem division". In the system division, the environment is described, abd the input/output-channels are named by user-defined names. The division problem can be compared to usual programs; where the I/O- channels are called by their names, as defined in the system division.
The hardware- dependence is thus restricted to the system part.

PEARL : TASKING and SCHEDULING

- WHEN	Interrupt		- ACTIVATE	task
- AT	time		- TERMINATE	task
- AFTER	duration		- SUSPEND	
- ALL			- CONTINUE	task
- ALL UNTIL			- RESUME	
- ALL DURING			- PREVENT	task

- EXAMPLE -

a) AT 12:00 ALL 10 SEC UNTIL 13:00 ACTIVATE
b) WHEN IR ALL 5 MIN DURING 2 HRS ACTIVATE

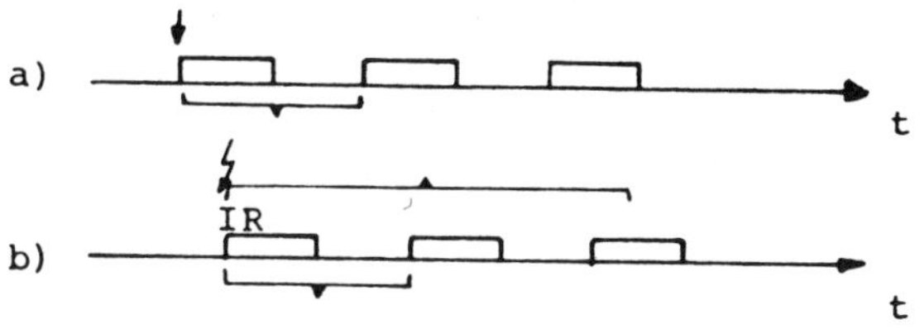

FIGURE 5

Of course, PEARL has the properties of other high-level
programming languages concerning arithmetic expressions and
program structures. These properties of PEARL are derived from
PL/1.
The system NCMES is designed as follows: The main task, which
is started by the operator, activates all the other tasks with
different priorities. According to the job the computer has to
fulfill, there are six tasks: two preliminary tasks, which
determine the probe- date and the position of the workpiece,
and four, which are resident during the measurement:

          - the task to control the machine
          - the task to read the next command
          - the task to print the results
          - the task to evaluate the measured data.

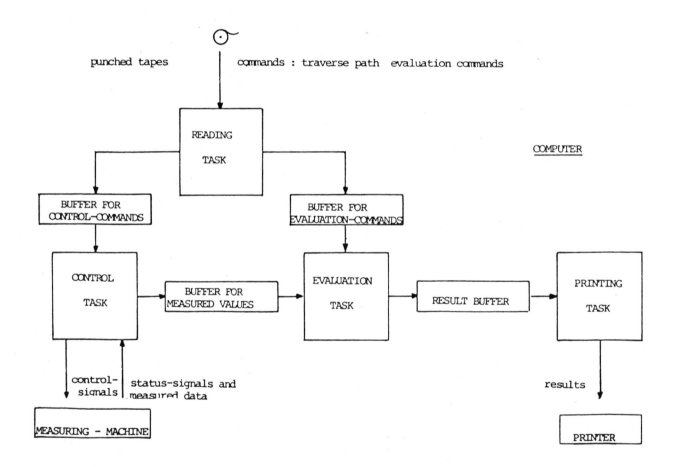

FIGURE 6

The control task has the highest priority, which means that if
this task is ready for processing, it immediately receives
service from the central processing unit, while the other
tasks go in waiting. Thus the machine is occupied. If this

task is not ready for processing - e.g. if it is waiting for the command buffer to be filled by the reading task or for the machine to reach the given position - the task with the next priority can be serviced. This is the reading task, which fills the command buffer. If the command buffer is not yet emptied or the task is waiting for an external device, the printing task may be run, and during its waiting time, the results can be computed.

It is ensured by synchronization-variables, that the tasks do not overlap each other, e.g. that the reading task has to wait, until the control task has executed all the given commands.

8. Conclusion

In this paper, a system for the control and evaluation of the measured data of the three-dimensional measuring machine has been presented. The system is programmed in the programming languages FORTRAN and PEARL, - the latter being more advantageous than the former. The comparision between the two systems leads to the result, that depending upon the type of measurement more or less great savings of the machine time are achieved, which increases the economy of the 3-D-measuring machine.

References

W. Eversheim   PROLAMAT 76, session IV
      et al.
                        "Programming of NC-Mearuring-Machines"

Terms:

Computers-programming, Coordinate Measuring Machines, metrology

Reprinted from Tooling & Production, April 1975

DIE SETS—

Plant layout and NC tools help meet the challenge

by **Robert B. Hurley**
Executive Vice President
Superior Die Set Corp.
Oak Creek, WI

To most people involved in metal stamping and tool and die making, the die set is as familiar as an old shoe, and about as exciting. But at Superior, the die set is the exciting and vital core of our business. Besides giving us the opportunity to serve a wide variety of businesses, the production of die presents some unique manufacturing challenges.

Where do we find the challenge in putting several holes in a couple of pieces of ground steel plate, inserting the appropriate guide pins and brushings, then checking to see that the punch holder moves on the guide pins in a satisfactory manner? The challenge lies primarily in the function of the die set itself.

As the foundation of a press tool system, it must be built to precise tolerances. The die builder must be able to depend upon it to provide proper tool alignment during die building, die maintenance and through the operating life of the complete die. The table on Page 52 lists a few of the applicable factors and tolerances that can be found in ANSI B5.25, 1968.

Our customers demand a wide range of die sets designed to meet the requirements for building and operating today's high-precision, high-production stamping tools. For instance, at Superior we offer 2 stock styles and 8 standard styles which are manufactured from over 60 different plate thicknesses. We build die sets ranging from 3" x 3" to 10 ft. and 20 tons.

Throw in the available guide pin and brushing selections (including ball bearing types), shank variations, bosses, thrust blocks, edge and handling selections and a myriad of other options, and you begin to see where we find the challeges in this business.

In addition, we are confronted with most of the problems of a jobbing shop: almost every order must be treated individually, and has its own unique processing and scheduling problems. Our customers ordinarily order only one die set at a time, and the orders arrive on a random basis. Only a small percentage of our sales volume is shipped from stock; therefore, work is of a highly varied nature.

Giving better, faster service depends on a number of factors. Two of the most important are plant layout and the use of numerically controlled machine tools.

Facility design

Production movement through the shops is crucial to meeting our goals and maintaining safe working conditions in the plant. Because we routinely handle workpieces up to 20 tons in an almost infinite variety of shapes and sizes, we try to handle and move the workpieces as little as possible. This factor was one of the most important considerations in our plant layout and in our decision to utilize the NC machining center concept.

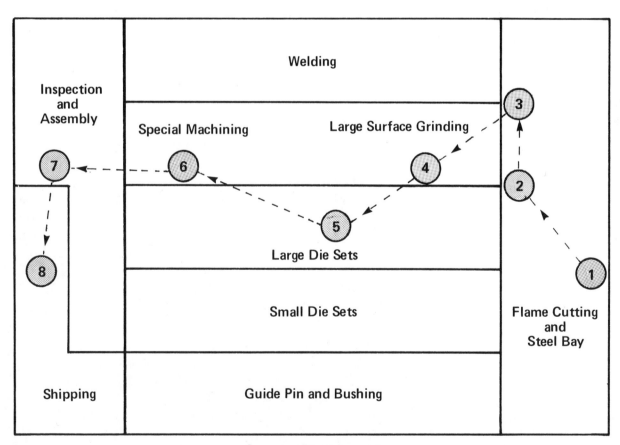

Superior Plant Layout

A typical die set production cycle will illustrate what I mean. I've chosen a custom die set since this is our specialty. Custom die sets typically require special flame cutting, welding, machining or other specific operations not ordinarily incorporated in the standard die set.

After the order entry process is completed, the order information is transmitted to our design department. Here NC flame cutting or machining tapes are prepared, special flame cutting drawings are made, where required, and welding and machining instructions are worked up. A cutting date is established based on the scheduled delivery date, and the amount of shop hours scheduled in the various work centers for the particular order.

This cutting date signals the initiation of shop activity on the major components of the order and controls release from the design department. In this instance we use the term cutting to mean the actual flame cutting of the die set components (punch holder and die shoe) from our steel plate inventory, which is in mill sizes. After flame cutting, the punch holder and die shoe are stress relieved, then moved to the Blanchard grinders for surface grinding.

After grinding, the components are moved either to one of our NC drilling and boring machines or to the special machine bay where horizontal boring bars, milling machines and radial drills perform a variety of boring, milling, drilling and tapping work. When economics or speed of operation dictate, some or all of the special machine work may be programmed and run on the NC machines. This decision is made prior to order release.

Foreman Harold Zierhart and Lead Inspector Dave Braun make a final quality control check on a custom die set. Die set is 16½" x 22", punch holder 6½" thick, die shoe 7¼" thick.

When all machining operations have been completed, the punch holder and die shoe are brought to the assembly and inspection area where final inspection is completed and the guide pins and brushings are assembled in the die shoe and punch holder. After final insepction, the set is moved to shipping.

The plant layout chart illustrates the movement of the custom die set. In general, the plant layout is designed for product movement through the work station sequences from one end of the plant to the other (right to left on the layout). In addition, certain machines are grouped in a production line to handle a particular size range of die sets. For instance, the small die set is geared to handle the heavier unit volume in smaller die sets (up to 20" x 20" and 400 lb).

Material movement in each production area is accomplished through the use of conveyors, overhead cranes, jib cranes and forklift trucks. The mix of these workhandling devices is determined by unit volume through the area and the weight and size of the product. For instance, in the small die set production area we primarily use roller conveyors to move the product from work station to work station and small jib cranes to move the workpiece from the conveyor to the machine.

Due to the large size of the work and a much lower unit volume, the primary mode of workpiece movement in the special machine bay is by overhead crane (1-20 ton, 1-10 ton) and by forklift truck. Individual work station cranes are also used to position the work on the machine.

Boring Bar Operator Louie Jakubczak prepares to finish-machine large cylinder bores in special die set plate. This part, 4½" x 60" x 128", illustrates the degree of machining effort that can be required in the production of special and custom die sets. Total die set weight was over 34,000 lb.

Typical Die Set Tolerances ANSI B.5.25				
		Tolerance		
Die Set Member	Characteristic	Commercial	Precision	
Punch Holder	Flatness	0.001″/lin ft	0.0005″/lin ft	
	Parallelism	0.0015″/lin ft	0.0007″/lin ft	
Guide Post	Diameter of working fit	+0.0005″ −0.0002″	+0.0003″ −0.0000″	
Guide Posts	Parallelism between guide posts	0.0015″ in 6″	0.001″ in 6″	

Shop work orders and job routing sheets are used to control product movement from one area to another. In practice, the movement is coordinated by the shop foremen who also control the loading at each machine in a particular machine group. The individual machine loading is dictated by the work schedule, the available work at the machine center, machine capability and operator skill level.

In addition, each foreman must take into account the tooling requirements, subsequent ooprations and scheduled hours on a job (here again we have a tremendous range, i.e., we have jobs in process requiring from less than 1 to over 500 bar hours).

Storage a problem

Storage of a workpiece before it is assigned to a machine in a particular machine center becomes a problem. Because we have a great deal of size variation in our product mix it is difficult to establish standard-size storage racks or a delineated storage area. Effective control of order releases and movement of product between machine centers minimizes storage problems and keeps jobs moving through the system. Here we rely on the routing sheets and the foremen, as well as order identification marks prominently placed on each workpiece, and a system of work skids where multiple pieces are involved on a single order.

Because of the trend to progressive dies, transfer dies and other more complex tooling, we have seen considerable change in the die set requirements of our customers during the last several years. Die sets have become more complex, with tighter tolerance specifications. This trend accelerated the expansion in our machining facilities. Fortunately, we initiated an analysis of our machining and other processing requirements just over five years ago. As a result of this analysis, we developed a program that has resulted in the purchase of a broad mix of conventional and numerically controlled machines. Included were Blanchard grinders; boring bars; NC drilling, milling and boring machines; NC and conventional flame-cutting and milling machines; a CNC lathe and a variety of other machine tools and accessories.

In our analysis and the subsequent economic justification we found several areas where NC controlled machines could increase our flexibility and efficiency. Our first effort concentrated on an activity that was common to all die sets--the drilling and boring of guide pin and brushing holes. At that time, die sets had to cross a least two machines to be drilled and bored. This required multiple setups and additional handling time in transferring the workpiece from one machine to another.

Our varying product mix with its wide range of drill and bore sizes dictated a tool-changer machine. We also wanted to incorporate the meximum amount of flexibility by selecting a machine with the largest table size and machine travel, within our budget range. Machine accuracy and repeatibility were important factors as were ease of programming and fixturing. Fixturing and the loading and unloading of the product from the machine were considered at the same time we evaluated the machine tool operating characteristics.

Our fisrt NC installation was a success and led to additional NC machines. We have installed an NC flame-cutting machine, a traveling column, NC tool-changing machining center with travel of 168" on the X axis and 72" on the Y axis, and have just started production with a CNC-controlled bar/chucker.

Tool use maximized

In all of our NC applications our goal is to maximize the amount of time the tool is doing work. This required dual fixtures on worktables, the design of master programs for generating NC tapes, special consideration of loading and unloading devices, and design and use of new tools. We had to consider the NC tool as part of our manufacturing system in order to take advantage of the speed, flexibility and accuracy offered by the numerically controlled machines that we have installed.

In the final analysis we are selling a service in the form of machine hours. Unfortunately, we can't take machine hours and put them on the shelf until one of our customers needs them. Once we have lost an hour through poor scheduling, a job error (yes, those do happen) or bad managemetn, it's gone forever, and so is the revenue that goes with it.

We have turned to NC machines in many areas to improve our overall productivity, to generate a greater degree of flexibility so that we can meet the present and future demands of our customers, and to achieve a better return on investment by reducing work in process, material handling costs and other indirect costs not often considered in NC justification procedures.

CHAPTER 3

CNC AND FLEXIBLE MANUFACTURING

Reprinted from Production and Inventory Management, Third Quarter, 1978

FACTORS TO BE CONSIDERED WHEN EVALUATING THE PURCHASE AND USE OF NUMERICALLY CONTROLLED MACHINE TOOLS

Henry A. Tombari
California State University   *Hayward, CA*

INTRODUCTION

The production of precision metal parts is one of the most expensive of manufacturing processes, yet cost reductions and productivity increases in this area have been slight.[1] This apparent lack of large productivity increases exists while many precision metal cutting and forming methods are on the market and being continually improved. One of the methods presently available that offers great productivity improvement potential is the numerical control (NC) machine tool. It has been estimated that numerical control machine tools represent only about 1% of the total machine tool population, about 21,000 out of a population of 2 million in 1972 and approximately 30,000 out of a population of 3 million in 1975.[3] Many of the NC tools are found in large batch operations, yet small batch manufacturing, the area in which NC is most applicable, accounts for 50% to 75% of the total national outlay for parts manufacturing.[4]

It appears that there are many manufacturing facilities that could benefit from the use of NC machine tools. Thus, the purpose of this paper is to examine the factors to be considered in the evaluation of purchase and use of NC machinery.

WHAT IS NUMERICAL CONTROL

Numerical control (NC) is a specialized form of automation. Specifically, automatic machine tools are programmed to perform tasks in accordance with numerical commands usually supplied from a punched paper tape.

NC fits into the spectrum of automation between conventional machine tools such as a lathe or drill press which are operated by skilled machinists, and machines fully automated to perform one specific task, such as automatic screw machines. NC machines are usually as expensive, but not always as fast, as fully automatic machines. NC machines are much more flexible, however, since they are directed by software (numerical commands) rather than hardware (gears and cams). While NC machines can make any part made on a conventional machine, the time and effort required to generate the NC program usually makes NC impractical for one of a kind items.

To obtain an NC program for a specific part, the programmer must first decide on a method of holding the raw material in place and pick a fixed reference point. Then he must transform the blueprint into rectangular coordinates and mathematical equations that completely describe the part in relation to the fixed reference point. The sequence of machining operations and the specific tools, feeds, and speeds must also be specified. All of this information is transformed into a series of numbers which become the commands for the machine. These commands are provided to the machine by a punched paper tape or directly from the memory of a digital computer.

Actual motion of the machine tool is accomplished by an electro-mechanical or electro-hydraulic servo system. Transducers located on the various moving axes of the machine provide an electrical signal indicating the position of each particular element of the machine. This electrical signal is provided to the control unit which compares the actual position with the position specified by the numerical command. If the two positions differ, the control unit provides a signal to the actuators (electric motors or hydraulic pistons) that move that machine tool element to the specified position.

In short, NC allows an automated operation of machine tools by such means as a system of electronic devices and changeable tapes. The electronic device interprets the programmed taped instructions and automatically directs the machine tool through the programmed sequence of operations while controlling machine machine speeds and feeds as well as other critical machine functions.

NC tools span a wide spectrum of capabilities. NC ranges from the relatively simple conventional NC, or hard-wired NC, controlled by punched paper tape to the computer numerical control which permits the input data to be introduced via a distant computer. Other applications of NC are found in machine techniques such as Computer-Aided Design—Computer-Aided Manufacturing (CAD/CAM) where the computer is used in all phases of manufacture from product design to production control. There are some great advantages in the use of NC in the proper situations, but there can also be some disadvantages.

ADVANTAGES OF NC

The primary advantage of NC is reduction of direct labor required per part. Many justifications for NC are based on this factor alone. To completely evaluate NC, however, many other advantages should be considered.

Set up time is considerably reduced, since the workpiece need only be positioned so that all dimensions are within the limits of machine travel. The machine then establishes a reference point and does the rest. No precision jigs or fixtures are required. Set up becomes merely a function of clamping the workpiece in place. In addition, if a NC machining center is used several operations are performed on the same machine eliminating the requirements for several set up operations and change-over to production of a new type of part requires mere insertion of a new tape.

NC machines spend no time checking their work, reading blueprints, or making calculations to determine speeds and feeds. As a result, they spend most of their time cutting, in fact about 75% of time is spend in cutting as compared to 18% for conventional machines.[5]

NC machines also usually produce high quality products. Moreover, the quality of parts are normally not affected by operator fatigue at the end of a shift or before the weekend.

Further, the use of NC equipment can usually reduce the overall operating costs of metal cutting and forming. Cost reductions partly result from the greater operating time available with NC machine tools versus conventional machining techniques. Moreover, due to advance planning and preparation that takes place in designing the system, NC equipment usually requires less debugging than conventional new equipment. For example, start up savings alone could have justified the capital investment associated with DNC at the Westinghouse plant at Round Rock, Texas.[6]

Another key advantage of NC is its flexibility. As mentioned above, changing over to produce a different part merely requires changing the paper tape or computer program. Design changes can be accomplished by changing the program which is particularly easy if the new part is similar to the old part. New jigs, fixtures, forging dies or cutting tools are not required. Hence, product improvements are easily implemented. Complex parts can often be machined from solid stock rather than castings or forgings that require extensive tooling. Additional tool savings are realized with continuous path machines which can often machine shapes with room space and inventory costs are reduced since a smaller selection of tools, jigs, and fixtures is required.

Additionally, accuracy of parts machined on NC equipment is a function of only one variable: machine response. Management controls this variable in the selection of equipment. Accuracy of conventional machining, however, is a function of several variables, such as operator judgment and skill, tooling, set up, and machine response. In the latter case, management has much less control over the variables.

Moreover, repeatability from part to part is excellent with NC machine tools. Variation from part to part tends to be minimized. Each part is machined exactly the same way with the same feed, speed, depth of cut, sequence of operations, etc. The only possible variable is tool wear.

Additionally, conventional machining requires many measurements and inspections to obtain an accurate piece. Properly maintained, NC machines produce accurate parts every time. Critical tolerance parts are just as easy to produce as parts with non-critical tolerance. Thus, NC machines are ideal for making interchangeable parts. The consistent dimensions of NC produced interchangeable parts can reduce assembly costs by insuring that all parts fit properly.

With regard to lowering costs, NC machines damage a part only with a major malfunction. Comparatively, the probability of product damage, particularly in machining to close tolerances, tends to be much higher with conventional machining. Thus, scrap losses are usually lower with NC.

Quality control is an additional advantage of NC, for inspection frequency can be drastically reduced. Inspection with NC should concentrate on the first part produced for all successive parts will be exactly the same within the given tolerance of the machine. Thus, only periodic checks as determined by statistical quality control are needed after the first part is checked. Inspection costs, therefore, tend to be considerably reduced.

Optimum machining conditions, including high speeds that require feed rates faster than a man could accurately control, are merely a matter of programming with NC. Further, cutting at optimum conditions requires less horse power, results in a longer and more predictable tool life, and produces a better surface finish, thereby reducing the number of finishing operations required.

Floor space can also be conserved through the use of NC. NC machines don't have the handwheels and other appendages required for manual control and are usually as small or smaller than conventional machines. Since a machining center replaces several machine tools and paper tapes replace jigs and fixtures required to produce various parts, floor space reductions of up to 70% are not unrealistic.[7]

As well as conserving space, inventory costs can also be reduced in several additional ways by NC. For example, smaller batch sizes and faster production result in less work-in-progress inventory. Since parts can be made quickly if required, a smaller inventory of finished products is needed. Accurate prediction of production time further allows closer alignment of raw material demands with production schedules, hence a smaller inventory of raw materials is required.

Overall by performing several operations at one machine, materials and handling costs tend to be reduced.

NC machines thus can offer their owners a competitive advantage. Due to the fact that several operations are condensed into one and machining time is very consistent, accurate production planning and forecasting can be done. On competitive jobs, the manager can bid very close to his costs because he can accurately determine his costs thus reducing the uncertanties of quoting a price in relation to costs. Additionally, product design and improvement need not be limited by tooling, machine configuration, or tolerance limitations. Design becomes merely a case of specifying what is desired.

Finally, skilled labor requirements are reduced with NC. Skilled labor, in addition to being at time difficult to find as well as expensive, is also usually concentrated in industrialized cities. Thus, freedom from skilled labor requirements allows location of facilities in smaller towns where real estate, taxes, and cost of living could be less expensive.

In sum, NC offers the user a variety of ways to reduce costs and increase productivity. Some of these advantages may occur in the short-run whereas others may be long run considerations. NC, however, is not a panacea, there are limitations and some highly undesirable disadvantages in their use. In the utilization of NC machine tools, the design of the system must investigate all the evidence that bears on the subject and thus present both the advantages and disadvantages.

DISADVANTAGES OF NC

The primary disadvantage of NC equipment is high cost. The increased fixed cost of an NC machine requires careful planning and accurate forecasts of sufficient production volume. Machine tools cannot be laid off if production requirements suddenly drop!

NC machines can cause financial disaster if management fails to adjust its thinking to exploit the advantages of quick set up, short machining time, and flexibility. To keep NC tools busy, jobs must be planned in advance and background preparation (programming, material purchasing, etc.) fully developed before the machine becomes free. This kind of planning is usually time consuming and is not always possible in a job shop operation.

Choosing personnel to operate, program, and maintain NC equipment requires discretion. The work force must be prepared and trained for the introduction of NC and accept this new innovation. Operators need to be observant and willing to learn the function of each detail of the machine.

Further, computer programmers will probably be needed in the operation. Programmers need to be adept at manipulating numbers and thinking abstractly. If possible, it would be helpful if they have a background in conventional machining.

Additionally, maintenance personnel should be "system" oriented. If possible maintenance mechanics should have some background in electronics, hydraulics, and mechanisms. The typical maintenance man, who is accustomed to trouble shooting mechanical systems primarily by "eyeball" inspection, will probably have to be trained to maintain NC equipment. Trouble shooting NC requires a logical systematic, step-by-step, process of elimination to locate one faulty electronic component out of thousands of possible offenders. Selection and training of maintenance personnel is of great importance and at times difficult. For example, of 80 men put through a series of training exercises at the Caterpiller Tractor Company, only 9 were chosen to be NC maintenance men.[8]

The problem of maintenance is accentuated with NC. Breakdowns are not common, hence diagnosis based on experience is almost impossible for the individual user. Yet the lost revenue and production delay that results from breakdown of an NC machine is much more serious than breakdown of a conventional machine.

In addition, one of the weakest points in an NC system is the tape reader. Without proper preventative maintenance, a tape reader can lead to a high frequency of NC malfunctions. However, certain NC techniques do limit the tape reader problem. For instance, direct numerical control (DNC) is designed to reduce the dependence on the tape reader in the machining operation.

It has been estimated that 50% of all NC installations are not economically successful.[9] This estimate implies that the other 50% are accomplishing their economic objectives. Thus, in weighing the possible advantages and disadvantages of NC machine tools in the evaluation of their use, the manager must view the specific manufacturing situation in which he finds himself. The manager should weigh all the alternatives, and include such factors as the "state of the art" in NC and the competition's position before deciding on the need for NC.

PLANNING FOR NC

Conversion of any manufacturing operation from conventional machining to NC is more than a change of equipment, it is a change of system and its concomitant philosophy. Many of the advantages of NC noted in the previous section are based as much on good planning and management as they are on unique properties of NC machinery. It would seem reasonable to assume that many of the unsuccessful NC applications could have been improved by better planning.[10]

The first step in NC planning should be the appointment of an NC coordinator. This person would insure that all factors impacting of the NC decision are evaluated. Specialized training, such as the one-week course offered by John A. Moorhead Associates, and sponsored by the National Machine Tool Builder's Association, is advisable for the NC coordinator selected.[11]

Within the planning parameters for NC system, consideration should be given to NC applications, capital investment and personnel factors. The following list of planning factors outlines some of the most important variables, but is far from being all inclusive. This list should be considered as only a starting point in making an evaluation for the purchase and use of NC machine tools.

APPLICATION FACTORS

One of the first decisions to be made is selection of products or parts to be machined using NC. The governing factor is batch size and demand. Ideally, there should be several different parts with intermittent or cyclic demand to justify NC. Other features of a machine part that would make it a good candidate for NC are:
1) At least some complexity.
2) Multiple machining operations or machining of more than one surface required.
3) Frequent design changes or new models.
4) Moderate to tight control of tolerances.

5) A high set up time.
6) Requirement for jigs and fixtures.
7) New parts for which no tooling exists.
8) Operator skill and attention required.

The point of exploiting the flexibility of NC cannot be over emphasized. If production control, inventory control, manufacturing engineering and other production functions utilize NC in the same manner as conventional machining, the full advantage of NC will not be realized. Therefore, it is imperative that everyone from top management down should fully understand the system philosophy of NC.

Other possible application questions that can be considered are:

Is reduction of lead time beneficial? If so, how much?

Will reduction of inventory be practical if setup and changeover time are reduced?

Is floor space at a premium?

Will management allow 2 or 3 shift operations?

Any other consideration, such as checking the bearing capacity of the soil and foundation of the plant suggested by the machine tool builder.

Though the above list is not exhaustive, it gives a framework from which the planning for NC applications can be considered. However, from this type of applications factors list, a master plan or PERT chart should be established to coordinate and evaluate cost factors, selection and training of personnel, preparation of tapes, changes of layout, and other preparation before a machine arrives. To insure the best possible results on production of the parts which will be used to evaluate equipment performance, it may be advisable to have the machine tool builder prepare the tapes and guarantee production rate, accuracy, and changeover time as part of the total NC cost.

In considering applications care must be taken to include all benefits, such as floor space savings and reduced in-plant transportation costs. In addition, some quality, reduced lead time, and improved production flexibility tradeoffs should also be considered.

CAPITAL INVESTMENT FACTORS

A fully developed economic justification is a necessity for the capital investment of the size necessary for NC machine tools. Many of the benefits of NC are difficult to quantify in terms of a dollar value. Economic evaluation of the first NC installation is particularly complex because of the difficulty in anticipating and economically evaluating all benefits that may result. A certain allowance for "manufacturing research" should be made for the first NC purchase.

Economic justification is a two step process. First, one must establish what future savings or earnings flows will be over time, and second, it must be decided if that earnings flow or savings are high enough to justify the investment. There are numerous techniques and formulas for accomplishing the second task but the first step is more difficult and requires expert attention.

Of the several methods for evaluating machine tool capital investment, the MAPI (Machinery and Allied Products Institute) and discounted cash flow methods appear to be the most popular. For example, at Nu Mac, the all-NC plant of the Ex-Cel-O Corp., Troy Michigan, the MAPI method was used.[12] However, in evaluating the capital investment in NC the cash flow method does offer some advantages.[13] Whatever investment evaluation technique is selected, objective consideration should be given to all elements that influence the decision.

PERSONNEL FACTORS

The cooperation of the employees on the plant floor is essential for the success of any new machine adoption, NC or conventional. Union leaders and workers should be informed in advance about the planned introduction of a new NC machine. Explanation of what NC can and cannot do should be presented to all who will be effected long before the machine is on the plant floor. Where possible, employees should be given an opportunity to contribute ideas to the introduction of new NC equipment and to participate in decisions that will effect their work spheres.

The introduction of NC also offers a chance to upgrade many jobs in the plant such as the introduction of programmers and sophisticated maintenance jobs. If possible, "in house" employees should be trained for the new technology. Overall, NC programming and maintenance positions provide an opportunity for increased technological challenge and promotion of the present workforce if the change is well planned.

Selection of programmers, operators, and maintenance personnel depends on the size of the work force available. In a small shop, all three functions could be performed by the same person. However, best advantage of NC is made when these functions are separated.

A good mathematical aptitude is essential for programmers. A high school education in most cases, is sufficient for occupational basic programming. However, for full time programming responsibilities of complex parts, a college education may be required. For example, Northern Illinois University offers a B.S. degree in computer sciences which includes a parts programming speciality.[14] But an individual with 2 years of technical school and experience in conventional machining can usually satisfy NC programming requirements.

Operators, on the other hand, need not have a conventional machining background or a high formal education. They should be intelligent, perceptive, and have an understanding of each function the machine performs. For an operator who thoroughly understands how the machine functions can be very helpful when "trouble shooting" is required on the machine.

Some companies prefer to put their highly experienced personnel on the most expensive machine. For instance, the General Machine Shop of Corning Glass encourages machinists to bid on new NC machines. This firm then selects the senior qualified bidders as the NC operators.[15] Other shops, such as the Tri-Kris Company in Lansdale, Pa., have done well using operators with little or no experience.[16]

As mentioned previously, maintenance personnel usually require knowledge in electronics, mechanisms, and sometimes hydraulics. Choice of maintenance personnel must be very selective to obtain the necessary skills. It should be emphasized that the success or failure of the NC operation may well depend on the proper selection and training of the maintenance personnel.[17]

Initial training of NC operators and programmers can usually be obtained from the machine tool builder. If needed, additional operators can then be trained in-house. Programmers may also receive additional training from computer service bureaus if time sharing facilities are used. Background training for all personnel can usually be obtained at a local community college, although some companies offer in-plant training. For instance, Clark Equipment Company, offered a 3-year NC maintenance course that met three hours per week at night.[18]

CONCLUSIONS

In summary, there appears to be no simple solutions in the evaluation of factors which have impact upon the decisions concerning the purchase and use of NC machine tools. Each situation must be thoroughly analyzed and objectively evaluated to determine the salient factors which will lead to a successful NC installation and operation.

From the very inception of the NC evaluation study to its final decision, top management should be totally involved. Top management must support a philosophy and give policy direction to a plan that permits the NC coordinator, and all others involved, the necessary time to objectively collect the necessary data and present well documented NC alternatives. It then is up to top management to decide the "best" alternative and support the implementation of that decision, if the NC purchase and use is to be a profitable success.

[1]Frank Hennig, "Why Buy an NC Machining Center?" *Automation,* September, 1973, p. 82.

[2]"A Landmark Test for DNC Machine Tools," *Business Week,* May 27, 1972, p. 36.

[3]James B. Pond, "Manual Machine Future Aided by NC Concepts," *Iron Age,* June 2, 1975, p. 37.

[4]Nathan H. Cook, "Computer-Managed Parts Manufacture," *Science American,* February, 1975, p. 25.

[5]S. Peter Kaprielyan, "NC and Small Business," *Instruments and Control Systems,* April, 1972, p. 21.

[6]Robert A. Wilson, "Bridges to the New Manufacturing Systems," *Iron Age,* August 19, 1974, p. 67.

[7]Frank Hennig, "Why Buy an NC Machining Center?" *Automation,* September, 1973, p. 83.

[8]Brian D. Wakefield, "NC is Really A-OK if Maintainers Know Job," *Iron Age,* January 25, 1973, p. 41.

[9]John H. Greening, "Build a 'Master Plan' for NC," *American Machinist,* March 22, 1971, p. 71.

[10]Frank W. Wilson, *Numerical Control in Manufacturing* (New York: McGraw-Hill Book Company, 1963), pp. 10–11.

[11]Robert A. Wilson, "Thinking NC Tools? First Think Training," *Iron Age,* December 2, 1974, p. 41.

[12]Frank W. Wilson, *Numerical Control in Manufacturing* (New Yrok: McGraw-Hill Book Company, 1963), p. 248.

[13]Harold Bierman, Jr. and Seymour Smidt, *The Capital Budgeting Decision* (New York: The Macmillan Company, 1965).

[14]Brian D. Wakefield, "Wanted: NC Maintanance Men—Good Ones," *Iron Age,* April 12, 1973, p. 48.

[15]James F. Gaertner, "Selling NC to the Workers," *American Machinist,* March 22, 1971, pp. 69–70.

[16]Robert A. Wilson, ::Low-Cost NC Lathe Can Answer the Skilled Operator Shortage," *Iron Age,* July 19, 1973, p. 47.

[17]Brian D. Wakefield, "NC is Really A-OK if Maintainers Know Job", *Iron Age,* January 25, 1973, p. 41.

[18]Wakefield, "Wanted: NC Maintenance Men—Good Ones".

About the Author—

HENRY A. TOMBARI is Assistant Professor of Management Sciences, California State University, Hayward, CA. He is a graduate of Rensselaer Polytechnic Institute in Civil Engineering, and he received his MS in Management (OR) from the United States Naval Postgraduate School and his DBA from the University of Maryland. His research and teaching focus on such varied areas as Production Management, Entrepreneurship, R&D Management, and Business Strategy. He is currently researching and writing a book concerning the sociotechnical approach to production management. He is a consultant to numerous organizations. He is a professional engineer with over twenty years of experience in engineering, facilities and construction management. His articles have been published in *Production Management* and the *Journal of Maritime Law and Commerce*.

The operator just loads and unloads the line as CNC guides four different parts through a three-machine "complex" at Caterpillar Tractor Co.

CNC Flexibility Accommodates 'Family of Four' In Multi-Machine Line

Requirements for these four parts are like those for many of the parts produced at the Aurora plant. Production requirements are in the mid-volume range. Machining tolerances are not excessive; scrap rates must be minimal. The family of parts must be run on the same machine to get maximum equipment utilization. Changeover from one part to another must be simple, with minimum tool and fixture changes. And, for best labor utilization, machine cycling should be automatic.

A new machine "complex" is meeting all these requirements with the aid of some reconsiderations in both the cutting tools and the coolant.

The parts are machined from ductile iron castings. About 21.5 in. OD, the unmachined parts weigh between 45 and 60 lb. Some 20 to 30 percent of that weight comes off in the form of chips during two chuckings and a drilling operation.

Machine Concept. The machining operations on these parts are being done on a complex of three machines supplied by Motch & Merryweather Co. An important factor in this development was the decision to use dual-spindle chucking machines for the primary metalcutting operations so two parts could be processed at a time. This gives a high degree of efficiency.

The final drilling operations are performed on a multiple-spindle machine, with the drilling time less than half that of either of the chucking cycles.

The Motch line uses two Twin-Spindle 225 VNC vertical chucking machines, plus a Motch vertical drilling machine. The three machines, caliper-type loading and unloading stations, and two turnover stations all are under the control of two Allen-Bradley 7360 CNC systems.

An operator takes these brake piston castings out of tote boxes and places them on the gravity roller conveyor at center. They feed down into position for the first chucking operation. Finished parts are delivered back to the operator from the drill machine at right

The operator loads the parts on a gravity roller conveyor. They are then power conveyed into position ahead of the first chucking machine. At that point, the lift-and-carry, load/unload mechanism picks up a pair of machined parts from the chuck jaws and a pair of unmachined parts from the conveyor. It then transfers the machined parts out of the machine and lowers them onto a following conveyor while simultaneously bringing the two rough castings into the machine and lowering them onto the chucks. The spindles on the two chuckers all have spindle positioning so the chuck jaws never interfere with the transfer grippers.

Machined parts then are moved on a conveyor through a part-turnover station and are positioned prior to the second chucking operation. Room has been provided so that a gaging station could be inserted at this point to check operations performed in the first chucking and transmit the necessary information to the CNC control for part and tool tolerance updating.

A similar lift-and-carry transfer system unloads and loads the second chucking machine. The transfer section between the second chucking station and the Motch drilling machine is a nonsynchronous conveyor which provides parts-in-process storage. Again, space has been provided for an optional gage station. This section of the conveyor also is arranged with a turnover unit. Some of the different parts must be inverted prior to drilling; the others simply pass through the turnover unit without action being taken.

Parts are processed through the drilling machine one at a time, since the drilling times are a fraction of chucking cycle times. This machine is controlled by a solid-state programmable controller.

Pistons come out of the first chucker at right, around the conveyor and into a turnover device, positioning them for Chucker No. 2 at left. They then go to the drilling machine at center rear

Two pistons, inverted after the first chucking operation, are loaded into the chucks for the second turning and boring sequences. For photographic purposes, the spindles have been stopped and the guards removed

After the parts have been drilled, they are ejected onto a gravity roller conveyor on which they move back to the load/unload station for unloading. The total machine complex, including the conveyors, is about 30-ft wide and 60-ft long.

Tooling Changes. Nearly all of the tooling for the two 225 VNC chuckers consists of Motch standard dial adjustable boring and turning bars. Dorsey Sisler, Caterpillar staff engineer, Manufacturing Engineering Group, says: "At first, we used titanium-coated carbide cutters. Our roughing cuts are $3/16$ to $1/4$-in. deep; we rough at 600 sfpm, and we finish at 720 sfpm. We tried different cutting tool materials until we found we could get good tool life (about double that of the titanium-coated tools) with Kennametal's KC210 nitrite-coated tools.

"We also have made a change in coolant, going to a Quaker Micro-Cut grade which has a lower nitrite content than our conventional water-soluble grades."

As for changeover flexibility, Sisler says no tool change is required for the chucker turrets, but that the chuck jaws and the fingers for the load/unload handlers must be changed. Changes in all three machines plus the entry of a different program in the controls can all be done in a shift, Sisler explains.

By tying the machine tools and all of the handling together in a CNC complex or system, Caterpillar has eliminated separate setups in standard machines, with separate queues, separate controls, and separate operators. Production efficiency on this family of four parts is thus at a maximum. □
Robert F. Huber

Justifying DNC

By Jack Sim
Field Sales Manager
Sundstrand Machine Tool

Necessity is the mother of invention. The need for flexibility in manufacturing brought about numerical control. The need for improving NC productivity brought about DNC. However, the cost disciplines in every company today require a justification to guarantee a return on a DNC investment. This paper will identify the major efficiencies in DNC in shops using one through ten NC machines. The conservative numbers used in this paper indicate that DNC will provide a payback for the majority of NC users.

1) Information Sources

2) DNC Costs

3) Formula Inputs

4) Formula Results

5) Conclusion

1) Information Sources

Sundstrand introduced a DNC system in 1968 and now has 19 installations in operation. A recent survey was made of five of these installations to evaluate actual benefits. The figures used in our justification formula are a conservative average from this survey.

Sundstrand - Denver, Colorado
A manufacturing plant in the Advanced Technology Division of Sundstrand Corporation. The NC operation in this plant can be considered a subcontract shop. While part shape, material, and lot size vary continually, this shop is a perfect example of how to operate NC machines in a production environment.

5 - Sundstrand OM-2 Omnimils	4 Axis
2 - Sundstrand Model 21 Omnimils	3 Axis
1 - Sundstrand 14NC Omnilathe	2 Axis

Hughes Aerospace - Tucson, Arizona
The plant manufactures missiles and has been a leader in the use of NC machines for many years.

3 - Brown & Sharpe Hydrocenters	3 Axis
1 - Sundstrand OM-2 Omnimil	4 Axis
1 - Kearney & Trecker Turn 12	2 Axis
4 - Hughes MT3 Machining Centers	3 Axis
1 - LeBlond Lathe	2 Axis

1) Information Sources (continued)

Heintz - Philadelphia, Pennsylvania
This manufacturer is basically a subcontractor on aircraft jet
engine parts. The biggest DNC benefit identified by this company
(but not used in our formula) is the savings in lead time between
plan and performance on their DNC equipment.

2 - Sundstrand OM-3 Omnimils	5 Axis
1 - Bullard VTL	2 Axis

Fairfield Manufacturing Company - Lafayette, Indiana
Fairfield manufactures shafts, gears, and gear box systems for
all types of industry. They are chip removal experts and confirm
the productivity benefits of DNC by operating two independent DNC
systems -- due to the location of two separate plants.

System I

2 - Warner & Swasey SC-28's	2 Axis
2 - Cinti Turning Centers	2 Axis
4 - Sundstrand 400S Omnilathe	2 Axis
4 - Sundstrand 400C Omnilathe	2 Axis

System II

2 - Giddings & Lewis Numericenter 10V	3 Axis
1 - Giddings & Lewis 25H	3 Axis
1 - Giddings & Lewis VTL	2 Axis
2 - Sundstrand 400C Omnilathe	2 Axis
1 - Warner & Swasey SC-28	2 Axis
1 - Sundstrand OM-3 Omnimil	5 Axis

2) DNC Costs

The costs are charted for shops with one through ten machines. The
total at the bottom of each column is what a Sundstrand system would
cost for the number of machines at the top. The equipment is described on the left side of the chart.

In order to show how the rate of return varies (depending upon variations in operating needs between companies), the formulas were run
four times as follows.

1. Once for two (2) shifts and one (1) new program on each machine
    every three weeks.

2. Once for three (3) shifts and one (1) new program on each
    machine every three weeks.

3. Once for two (2) shifts and one (1) new program on each machine
    every week.

4. Once for three (3) shifts and one (1) new program on each
    machine every week.

	1 Machine	2 Machines	3 Machines	4 Machines	5 Machines	6 Machines	7 Machines	8 Machines	9 Machines	10 Machines
DNC Computer & CRT w/Keyboard	$80,000	$80,000	$80,000	$80,000	$80,000	$80,000	$80,000	$80,000	$80,000	$80,000
Additional Core					$ 4,000	$ 4,000	$ 4,000	$ 4,000	$ 8,000	$ 8,000
SPLIT Processor	$ 3,500	$ 7,000	$10,500	$14,000	$17,500	$21,000	$24,500	$28,000	$31,500	$35,000
Super SPLIT	$ 1,300	$ 1,300	$ 1,300	$ 1,300	$ 1,300	$ 1,300	$ 1,300	$ 1,300	$ 1,300	$ 1,300
Super SPLIT for Each Processor	$ 100	$ 200	$ 300	$ 400	$ 500	$ 600	$ 700	$ 800	$ 900	$ 1,000
Additional Disk Storage					$ 5,304	$ 5,304	$ 5,304	$ 5,304	$ 5,304	$ 5,304
CRT w/Keyboard & BTR Connection to Each Control	$10,500	$21,000	$31,500	$42,000	$52,500	$63,000	$73,500	$84,000	$94,500	$105,000
100 feet Cable	$ 58	$ 116	$ 174	$ 232	$ 290	$ 348	$ 406	$ 464	$ 522	$ 580
100 feet Cable	$ 58	$ 116	$ 174	$ 232	$ 290	$ 348	$ 406	$ 464	$ 522	$ 580
Cable Connectors	$ 25	$ 50	$ 75	$ 100	$ 125	$ 150	$ 175	$ 200	$ 225	$ 250
Cable Connectors	$ 25	$ 50	$ 75	$ 100	$ 125	$ 150	$ 175	$ 200	$ 225	$ 250
Totals	$95,566	$109,832	$124,098	$138,364	$161,934	$176,200	$190,466	$204,732	$222,998	$237,264

3) Formula Inputs

The formula we use is the most frequently used justification method per a survey taken by Metalworking Magazine. It is a combination of payback and rate of return based upon after-tax savings. Rate of return is the most important figure in that it represents an index of potential income generation. Payback measures the degree of risk in recovering an investment.

The following inputs are required in the formula, and the value is explained to make the justification as fair as possible.

1. Thirty minutes was selected as the cycle time per setup.

2. Twenty parts in a lot size were selected -- resulting in a ten-hour run per setup.

3. One minute was assigned to calling up a DNC program versus three minutes to load and unload a tape program.

4. Three hours were used to debug a new part program with DNC, and ten hours were assumed to debug a new tape program without DNC.

5. DNC programs run conservatively 15 per cent faster than tape programs per our survey of DNC users.

6. Tape costs of $5 per month per machine were eliminated with DNC when one new part program was made every three weeks. $15 was saved per month per machine when a new part program is made every week.

7. DNC machines had 5 per cent more uptime than tape machines per our survey of DNC users (but was not used in formula). This was primarily due to tape reader maintenance.

8. DNC provides a savings for in-house computer programming over manual or a terminal service. A $500 savings per month was used for this benefit when one new part program was made every three weeks. This savings increased to $1,500 per month when one new part program was made every week.

9. Operator pay was selected at $7.00 per hour. Programmer pay was selected at $8.00 per hour. No overhead was used in this justification exercise.

10. We used 2,000 hours for hours each machine was available per shift.

11. Corporate tax rate used was 48 per cent. Investment tax credit used was 7 per cent.

12. Investment life was 4, 8, 12, or 16-year periods.

4) <u>Formula Results</u>

The computer calculations are summarized on the following four print-outs. They are presented in the order of lowest return on investment first and highest return on investment last.

A. The "1" column lists the equipment cost of the ten DNC systems from the first chart.

B. The "9" column lists the dollar savings of the DNC method over tape. This figure is a simple calculation of hours saved multiplied by the operator or operator and programmer costs. For example --

<u>Tape - One Machine</u>

    17 new part programs to debug per year
    17 x 10 hours per program = 170
    170 hours x $15   = $ 2,550.

    400 tape changes for lot runs per year
     17 tape changes for debugging per year
    417 x 3 minutes per tape change
    21 hours x $7   = $ 147.

    191 hours from 4,000 hours available
    in two shifts = 3,809 production hours

    3,809 hours x $7   = <u>$26,663.</u>
                                       $29,360.

<u>DNC - One Machine</u>

    17 new part programs to debug per year
    17 x 3 hours per program = 51
    51 hours x $15   = $ 765.

    400 part programs for lot runs per year
     17 part programs for debugging per year
    417 x 1 minute per part program call-up
    7 hours x $7   = $ 49.

    3,809 hours production from tape system
    becomes 3,239 hours because DNC is 15% better
    on production.

    3,239 hours x $7   = <u>$22,673.</u>
                                      $23,487.

  $29,360.
- <u>$23,487.</u>
          $ 5,873. - machine savings
     + $ 6,000. - computer terminal savings ($500 per mo. x 12)
     + $ <u>60.</u> - tape savings ($5 per month x 12)

      $11,933. - DNC savings over tape for one machine
                per year on two-shift basis when
                one new part program is debugged
                every three weeks.

4) <u>Formula Results</u>   (continued)

    C. The "8" column designates the variations in years of machine
        life.

                8A - 12 years
                8B - 16 years
                8C -   8 years
                8D -   4 years

    D. The minimum return on investment for any company can be
        easily related to net earnings plus depreciation per year
        divided by fixed assets. If we assume this figure to be
        15 per cent, then we can justify DNC on even one (1) machine
        when the programming requirement is one new part program
        every week on either a two-shift or three-shift basis.

        The printout of formula results was limited on two or more
        machines to a machine life of twelve years, since we assume
        that to be normal for NC machines.

        The big difference in achieving a better rate of return
        on any machine combination and with a two-shift operation
        is the difference between new part programs occurring
        once every three weeks or once every week.

TWO SHIFTS   ONE NEW PROGRAM IN THREE WEEKS

			Rate of Return In %	Payback In Years
1A	9A	8A	5	9
		8B	7	9
		8C	0	-
		8D	0	-
1B	9B	8A	13	6
1C	9C	8A	18	5
1D	9D	8A	22	4
1E	9E	8A	23	4
1F	9F	8A	26	4
1G	9G	8A	28	4
1H	9H	8A	29	4
1I	9I	8A	30	3
1J	9J	8A	32	3

TWO SHIFTS　　　ONE NEW PROGRAM EVERY WEEK

			Rate of Return In %	Payback In Years
1A	9A	8A	18	5
		8B	19	5
		8C	14	5
		8D	0	-
1B	9B	8A	31	3
1C	9C	8A	41	3
1D	9D	8A	48	2
1E	9E	8A	50	2
1F	9F	8A	55	2
1G	9G	8A	59	2
1H	9H	8A	62	2
1I	9I	8A	64	2
1J	9J	8A	67	2

THREE SHIFTS   ONE NEW PROGRAM IN THREE WEEKS

			Rate of Return In %	Payback In Years
1A	9A	8A	7	8
		8B	9	8
		8C	2	8
		8D	0	-
1B	9B	8A	16	5
1C	9C	8A	22	4
1D	9D	8A	26	4
1E	9E	8A	28	4
1F	9F	8A	30	3
1G	9G	8A	32	3
1H	9H	8A	34	3
1I	9I	8A	35	3
1J	9J	8A	37	3

THREE SHIFTS   ONE NEW PROGRAM EVERY WEEK

			Rate of Return In %	Payback In Years
1A	9A	8A	19	5
		8B	21	5
		8C	15	5
		8D	0	-
1B	9B	8A	33	3
1C	9C	8A	43	3
1D	9D	8A	51	2
1E	9E	8A	54	2
1F	9F	8A	59	2
1G	9G	8A	63	2
1H	9H	8A	67	2
1I	9I	8A	69	2
1J	9J	8A	71	2

5) Conclusion

An 18 per cent (12 year machine life) return on investment is available on one machine -- two-shift operation -- when a new part program is required every week. This rate of return is achievable on a justification method that does not use an overhead value in the calculation. It is also achievable when the differences between tape and DNC have been weighted on the conservative side.

The suspicions that we all have about the efficiency of the tape method we use in NC to transfer management directions to machines can now be confirmed by standard justification methods using actual results from DNC installations.

An investment in DNC is a worthwhile investment in obtaining the much-needed improvement in NC productivity.

Part programming in one-third the time, the elimination of paper tape, source language editing, and control over shop floor activity is why the use of DNC will double

Direct Numerical Control Works— Orders for At Least 50 New Systems Are Pending

DNC is alive and on the verge of staging a comeback. Future installations will be built around computer numerical control (CNC) systems. The market will increasingly shift away from the giant corporations and spread out more evenly among medium-size plants. The very large companies will turn more to designing, building, and supporting their own systems. And, DNC's ability to gather and generate management information will prove to be a key selling point.

These predictions were published two years ago in PRODUCTION. They have not been forgotten. Suppliers and users of DNC have been monitored. This is an update on what has occurred since April, 1977.

DNC IS ALIVE AND ON THE VERGE OF STAGING A COMEBACK

The market for DNC is healthier today than it has been in a long time. There currently are some very big buys up. Tom Shifo, general sales manager, White-Sundstrand Machine Tool, Inc., Belvidere, IL, said this two years ago. The past 24 months have proved his enthusiasm justified. Since April, 1977, orders for Omnicontrol systems have increased to 34 from 26. One of the new installations will be associated with a manufacturing system.

White-Sundstrand has sold all but about 20 of the DNC systems ordered to date. It remains optimistic on DNC's future. "The market will flourish as more people recognize the inefficiency of traditional methods for writing and debugging part programs and question less the use of electronics and computer technology. By the end of next year," predicts Shifo, "the number of DNC installations could increase to from 60 to 80 systems, up from 50 to 55 systems today."

Many builders now market DNC as part of a more encompassing manufacturing system. This has led to the coining of phrases like manufacturing systems, flexible manufacturing systems, distributed manufacturing systems, advanced batch manufacturing systems, real-time DNC, nonreal-time DNC, batch-mode DNC, and DNC-like systems.

Giddings & Lewis markets manufacturing systems under the name NumeriMation. It can provide machine tools, material handling systems like this dual-station pallet shuttle and DNC capability. One system has been sold.

Bob Chamberlain, vice president and general manager of Giddings & Lewis Electronics Co., Fond du Lac, WI, cuts through the confusion with these definitions: DNC controls several machine tools simultaneously from a remote supervisory computer which can also program and edit in source language. It is justified when the following conditions exist. 1. There is a very complex NC program involved and a high probability for program changes when cutting the first part. 2. When there is a communications problem, i.e., there is a significant cost in supplying and maintaining up-to-date part programs at respective work centers in coordination with manufacturing schedules. 3. Where a family of parts exist so that programs can be composed of subprograms without reprogramming.

A manufacturing system, says Chamberlain, consists of two or more machines which function as a coordinated unit under direction of an executive computer (CNC). Such a system can include integrated materials handling and be linked to a central computer.

Giddings & Lewis Electronics markets a manufacturing system. Chamberlain says DNC is not needed by most companies. It is too expensive when compared to the edit capabilities of today's CNC units. He places the return on investment for DNC at about 5 percent. ROI would be higher with a manufacturing system. To date, G&L Electronics has sold one manufacturing system.

Kearney & Trecker Corp., Milwaukee, sells its Gemini DNC package as a subsystem of its Flexible Manufacturing System. DNC represents about 5 percent of an FMS installation. Tom Klahorst, sales manager for K&T's Special Products Div. says it is to the user's advantage to view DNC as an element of a flexible manufacturing system. "As a rule of thumb, a stand-alone DNC line will improve machine tool utilization 4 or 5 percent. A manufacturing system, on the other hand, will result in from 8 to 10 percent better utilization."

Two years ago, Klahorst said it would be from 5 to 10 years before DNC came into its own as a standard product. Less two years, that timetable stands. "Within three years, I expect the number of manufacturing system installations to double to 100 or more. Within five years, that number could redouble again. Demand for stand-alone DNC systems will run at about four or five lines a year." For K&T, this renewed interest in DNC has resulted in two FMS sales since April, 1977. This brings to seven the number of lines sold.

Westinghouse Electric Corp. remains an active DNC supplier.

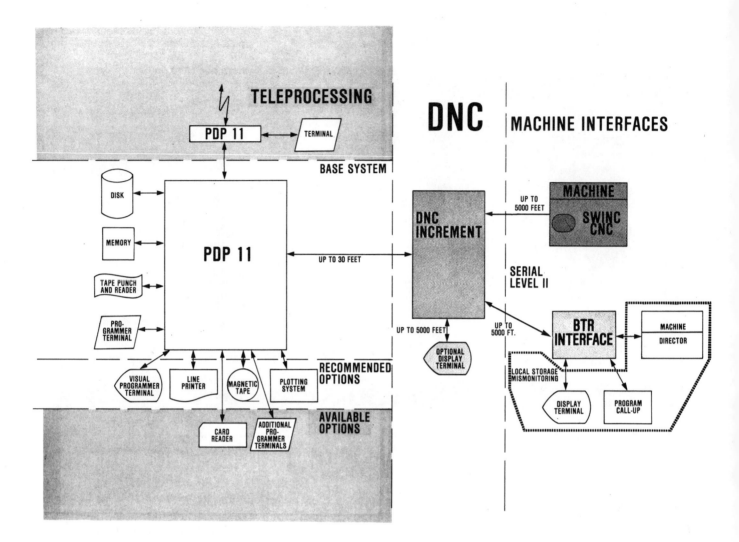

During the past two years, it installed a system to control from 25 to 30 machines at Northrop Corp. and is completing installation of a system which will control 64 machines at the Ft. Worth, TX, plant of General Dynamics.

Wayne Arnold of Westinghouse's Industry Systems Div., Pittsburgh, says: "There are from 20 to 25 DNC programs in some stage of active negotiations at this point in time. This is equivalent to half of everything installed to date." He adds that Westinghouse has quoted General Dynamics on two additional systems.

At Westinghouse, the biggest change to date is the use of a 32-byte super mini which allows full APT programming and comprehensive data base management. "Our experience," says Ar-nold, "indicates that simply down-loading programs was a breakthrough several years ago, but that the major increases in productivity now come from on-line collection and analysis of machine performance and in-process inventory. In fact, we have just completed a study for one customer which includes automatic storage and retrieval as well as distribution of material in addition to the DNC management information

---

"There are from 20 to 25 DNC programs in some stage of active negotiations at this point in time."

Wayne Arnold, Westinghouse

---

and part programming. We really feel that integrated manufacturing systems are here today."

Allen-Bradley's Systems Div., Highland Heights, OH, has three DNC lines to its credit: two at Boeing and one at Lockheed. Its present role in the marketplace is mainly one of supplying manufacturing system builders with programmable controllers for materials handling, numerical controls for .machine tools, and mini-computers for management information and supervisory purposes. It also builds a machine terminal which functions like a behind-the-tape reader to permit interfacing hardwired NC machines to DNC, and shortly will announce a package designed around a Digital Equipment Corp. PDP-1134 mini-computer to handle part program distribution, management information and tasks like line balancing.

Larry DeLong, manager of product marketing for the Systems Div., says A-B also plans to get more deeply involved with developing complete manufacturing systems. No lines have been sold yet, but A-B is working on a project.

Cincinnati Milacron, Inc., Cincinnati, a company which had machine tools under direct computer control in 1966, and demonstrated DNC at the 1970 Machine Tool Show, has never sold a pure DNC system. In fact, it no longer offers a separate DNC package. However, it does build multiple machine systems (Variable Mission Manufacturing) which can be interfaced with a Milacron DNC package as a subset.

Charles Carter, manager, Advance Machine Tool Systems, says Cincinnati plans to expand into manufacturing systems capable of handling a broader range of workpieces. Milacron systems sold to date are for limited workpiece variety. Carter is bullish on the future of DNC as part of a manufacturing system. In the past six months he has seen an upturn in market interest double that evident during the previous six months. "If a manufacturing system is defined as including at least eight new special purpose machines," says Carter, "the machine tool industry should be able to sell 20 new systems over the next two years. If the definition is broadened to include systems with fewer than eight machines and include retrofitting NC machines with CNC, this figure could go as high as 50 to 70 new systems."

General Electric Co. has changed its position on DNC. While it no longer markets its CommanDir line it is offering NC products that are directly compatible with DNC

systems. And its manufacturing plants are concentrating on improving and expanding DNC capabilities. For example, GE's 1050 CNC now offers a two-way communications option for downloading programs from a host computer and a new microprocessor-based part program storage and edit option called the μ-STOR Memory System which allows hardwired NC to be upgraded to CNC. Early this year, this package was updated to include a two-way communications options that permits programs to be down-loaded from a host computer. This permits a DNC link.

General Electric's Installation and Service Engineering Div. (INSE) serves as the company's focal point in responding to DNC customers. As the company's service arm for Mark Century NC controls, INSE also markets the μ-STOR Memory System to user customers. At present, INSE is developing custom interfaces to allow use of this system on non-GE controls thus permitting expansion of its DNC capability.

FUTURE DNC INSTALLATIONS WILL BE BUILT AROUND CNC SYSTEMS

This is no longer a prediction. It has come to pass. Today, instead of a host computer controlling machine tools on-line in direct fashion, programs can be downloaded to a CNC at the machine tool. The host computer takes on the role of a centralized data processing and program distribution system. The CNC stores the program and directs the machines. These Level II systems, by means of a behind-the-tape reader, make it possible to operate equipment on-line or independent of the central computer. This overcomes the objection of host computer dependence, and frees the more powerful host computer for other tasks.

The integration of CNC into the DNC environment has involved more companies in systems development and support. For example, the Actron Div. of McDonnell Douglas markets Actrion III CNC which interfaces nicely with DNC. McDonnell Douglas Automation

DNC is marketed by Kearney & Trecker as part of its Flexible Manufacturing System. This is the nerve center of the FMS installation at Rockwell

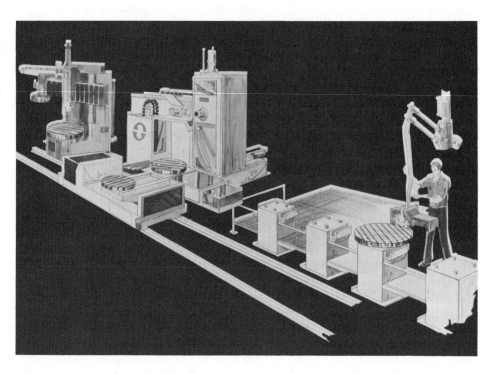

White-Sundstrand Machine Tool dominates the DNC field. It has sold 36 of the 50 to 55 domestic installations. Its Omnicontrol line is synonymous with DNC

Co. offers DNC/like services such as dial-up time sharing for parts programming and program storage on McDonnell's large computer. United Computing Corp., a subsidiary of McDonnell Douglas Automation, markets a part programming system that can be used to punch tape and feed a CNC. It also sells a minicomputer and minicomputer software package in effect, a mini-DNC system.

Vega Servo-Controls, Inc., Troy, MI, makes the use of CNC in the DNC environment attractive by offering a control which operates off a floppy disk memory and works in a higher language than conventional CNC's. The floppy disk provides bulk storage for part programs and inputs the control unit. It is being used with 24 DNC machines at Fairchild Republic Co.'s Farmingdale, NY, plant. It represents what is referred to as nonreal time DNC.

As CNC becomes more powerful, the configuration of DNC installations will continue to change. One day it may be possible to do source program editing with a CNC at the machine tool. This will enable small shops with three or four NC machines to cut programming costs. It will not be economical for companies employing or anticipating expansion to a larger number of NC machines.

THE DNC MARKET WILL INCREASINGLY SHIFT AWAY FROM THE GIANT CORPORATION AND SPREAD OUT MORE EVENLY AMONG MEDIUM SIZE PLANTS

This definitely is occurring. Many companies which discounted DNC as being too expensive a few years ago are reevaluating its use as hardware costs come down. The host computer for a small system now costs from $70,000 to $80,000. A larger system will run from $150,000 to $300,000. CNC machines can be tied into a system for about $3500 each. Conventional machines can be tied to a DNC line for from $12,000 to $15,000.

Tom Shifo, White-Sundstrand, believes that shops employing from 100 to 1000 persons is where the greatest market potential exists. Excellent examples of such companies are Fairfield Manufacturing Co., Inc., Lafayette, IN, and The Falk Corp., Milwaukee.

Fairfield Manufacturing has been involved with DNC since 1970 when it put an OM-3, two Cincinnati shaft lathes and three Warner & Swasey SC-28's under direct control of an IBM 360 computer. Today, according to Max Johnson, manager, Manufacturing Engineering, 19 machine tools are under direct control of a PDP-1120 computer, and another 42 NC machines get programming support from the host computer.

Fairfield manufactures gears and gear components in job lots averaging 300 pieces. Each week it generates from 25 to 35 new programs, 60 to 70 percent of which are for new part numbers. "The single greatest attribute of DNC," says Johnson, "is source language programming and editing. It permits us to support programming of 61 machines with five programmers."

Present plans at Fairfield call for replacing the PDP-1120 computer with a more powerful PDP-1134 and adding a White-Sundstrand OM-2 and one disk drive to the present one drive/one library setup. The entire DNC facility at Fairfield operates on a Level I basis. No programs are downloaded and stored at the CNC. In the future, however, Johnson says this is likely to change.

Falk Corp. operates DNC lines at two plants in the Milwaukee area. A total of 12 machines are under DNC. A company spokesman reports that there are plans to add two Cincinnati chuckers to the system and replace a tracer-type flame cutter with an NC unit which can be programmed by the host computer to perform part nesting.

Falk, which has a White-Sundstrand DNC system, generates about 40 new programs a week. Job lots range from 16 to 25 pieces up to 200 pieces. Like Fairfield Manufacturing, it views source language programming as the No. 1 attribute of DNC. Other companies which are successfully applying DNC today include the Louisville, KY, plant of International Harvester, Ingersoll-Rand's Roanoke, VA, plant, several divisions of General Motors, the Rock Island (IL) Arsenal and the Richmond, VA, plant of Western Electric. It is obvious

from this that DNC is no longer unique to aerospace firms like McDonnell Douglas, Hughes Aircraft, Lockheed Missile & Space and Pratt & Whitney.

THE VERY LARGE COMPANIES WILL TURN MORE AND MORE TO DESIGNING, BUILDING AND SUPPORTING THEIR OWN DNC SYSTEMS

This prediction of two years ago also has proved true. It is evident at General Electric, Westinghouse, Caterpillar and several aerospace companies. General Electric DNC installations are being upgraded or expanded at plants in Erie, PA; Evendale, OH; Hooksett, NJ; Schenectady, NY; Utica, NY; and Wilmington, NC. These efforts take precedence over efforts to market DNC outside of the company.

Westinghouse has in-house programs underway in at least four locations. At Round Rock, TX, where a DNC line was installed several years ago to machine gas turbine engines, a complete retrofit has been made to convert the line to producing large industrial motors. In all of these plants, a large portion of expenditures is earmarked for replacing NC with CNC units.

Stan Froyd, chief scientist, Actron Div., McDonnell Douglas Corp., Monrovia, CA, says the reason more and more companies are designing DNC lines themselves is that they want to tailor hardware and software to individual requirements. This is what McDonnell Douglas did at St. Louis and Torrence, CA, where a total of 150 machine tools are under DNC. The company considers DNC a key to its success in the aerospace field. This is why, says Froyd, that it is not particularly

anxious to package its expertise and offer it for sale.

THE ABILITY TO GATHER AND GENERATE MANAGEMENT INFORMATION WILL PROVE TO BE ONE OF DNC'S MOST VALUABLE ATTRIBUTES

Since April, 1977, management information systems (MIS) have come to be synonymous with manufacturing systems and DNC. Most suppliers offer packages to monitor manual machines, NC and CNC machines and shop personnel. All view MIS as being the heart of an effective installation and consider its absence from early installations a reason for DNC's slow-growing popularity.

Today, MIS is recognized by users as something which is needed. The only question which remains to be answered is whether it should be part of the DNC system, or be operated independently. This issue has more to do with labor relations than it does with implementation, and will most likely be dealt with by companies on a plant-by-plant basis.

Conclusion. It should be evident at this point that predictions made two years ago by suppliers and users of DNC have proved realistic. More important, it is evident from their comments that the acceptance of DNC as a stand-alone product or as part of a manufacturing system is accelerating. Key reasons continue to be the ability to reduce programming time by approximately one-third, eliminate tape, gain control over operations via management information systems and boost overall machine tool utilization. What remains to be seen is: will the doubling and redoubling of the number of systems installed materialize as predicted? *Brian D. Wakefield* □

Reprinted from the International Journal of Production Research, 1978, Volume 16 Number 6

Hardware and software developments for a DNC manufacturing cell

D. A. MILNER† and J. D. BRINDLEY†

Illustration is given here of the different development stages required in the retro-fitting of an existing numerical control machine, giving appropriate details of the interface link. Some software developments are indicated that enhance the range of the machine tool when connected on-line to a minicomputer, and mention is made of the production of Novikov gears.

Introduction

Many of the first and second generation of numerically controlled milling machines have been discarded, or at least are under-used, because they lack many of the facilities found on present machine tools being manufactured. These machines in some cases are in very good mechanical condition and would obtain a new lease of life as well as give a better return on capital if they were updated by means of a behind-the-tape reader DNC (direct numerical control) system.

With the reducing cost of the mini and micro computer and the increasing cost of machine tools, it is now becoming financially viable to couple the computer to some of the older NC (numerical control) machine tools, and thereby obtain the advantages of the latest technology.

In the Production Engineering Department of the University of Aston are several NC machine tools, some of which can be improved by the implementation of a DNC adaption. In 1973 a decision was made to purchase a Data General Ltd. Nova mini-computer in order to update a vertical milling machine to give valuable experience of DNC and adaptive control of machine tools. Work had previously been done using an Adaptive Control Constraint System (Milner 1974). The machine chosen for retrofitting was a $2\frac{1}{2}$ D Starrag vertical milling machine fitted with an Olivetti CNZ control system which incorporated a Tally tape reader, using EIA tape format. This machine was being used for the manufacture of prototype components, many of which were of complicated design.

The retrofitting and development of other such machine tools and peripherals would form a relatively small DNC manufacturing system, as shown diagrammatically in Figs. 1 and 2. The essential feature of this system is that it forms the nucleus of a manufacturing system with a computer and associated peripherals as integral and indispensible parts of the system.

The advantages to be gained from a behind-the-tape reader (BTR) configuration (Crossley and Lewis 1974, Inaba 1973) are:

    (i) flexibility, in that the system would accept data from the computer and tape reader;

Received 14 June 1976; revision received 17 July 1978.

† Department of Production Engineering, University of Aston in Birmingham, Birmingham, England.

Published by Taylor & Francis Ltd, 10–14 Macklin Street, London WC2B 5NF.

(ii) ease of part programming;

(iii) ability to type, check and edit part programmes;

(iv) programme storage facility, with quick access;

(v) cutter offset control;

(vi) English to metric conversion;

(vii) provision of subroutines, ' canned cycles ';

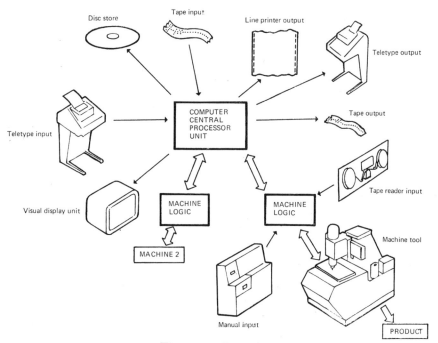

Figure 1. Data flow.

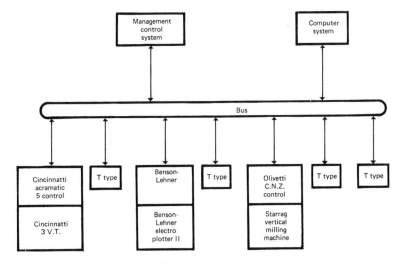

Figure 2. Cell.

(viii) circular interpolation.

DNC computer configuration

The general arrangement of the DNC system is shown in Fig. 3. The system uses a Data General Ltd. Nova 1200 series computer, which is configured through the DNC software to operate the system. The computer has 16 K of 16 bit core memory with moving head disc back-up store. Extension of the existing computer hardware can also be readily achieved to meet future needs.

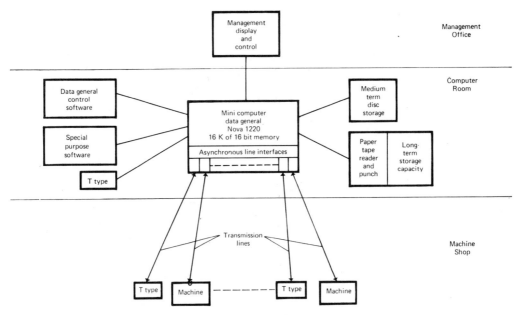

Figure 3. System configuration.

The DNC software is divided into a number of tasks. One task is responsible for communications with the machine tool operator and another is responsible for distributing part-programme data to the NC systems. The coordination is carried out by the software executive, it being the responsibility of the executive to ensure that when a high priority task needs to be carried out, it receives the necessary computer time without having to wait for completion of a low priority task. A time shared configuration gives this flexibility, and to achieve this a buffer store is added at the hardware interface to ensure continuity of information flow.

The transmission lines between the mini-computer and NC system may be about 1 km long. In order to minimize the number of cores required in the transmission lines, serial data techniques are used.

Interface components

Computer output board

A general purpose interface board was purchased from Data General Ltd. which fits into the main frame of the computer and has pre-wired input and output registers, together with all the necessary interrupt and device selection units. The memory and transmitter units that correspond with the BTR unit are also wired on this board.

Buffer store

The stores are 32 times 8 bit first out (FIFO) memory units with completely independent read and write operations. Several components are coupled together to give two storage units, each capable of handling 160 words.

Transmission link

The link between the computer and the system, as illustrated in Fig. 4, is a vital part and very high reliability of data transmission is essential for the success of any DNC system.

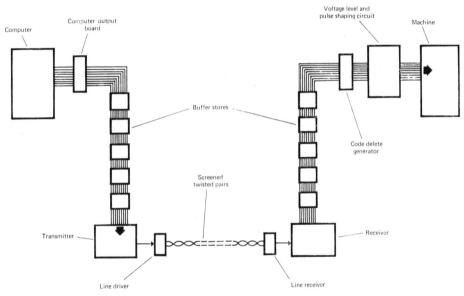

Figure 4. Interface.

The signals at the BTR unit and computer interface are of serial form and at levels conforming to an EIA standard. If the computer and NC system were close together, it would be feasible to link the BTR unit to the computer interface. In practice, the computer and NC system are widely separated and it is necessary to interpose some form of transmission system. Also the transmission lines use twisted pairs with an overall screen to minimize inductive and capacitive coupling.

The transmitter accepts characters from FIFO buffers in parallel form and transmits them in serial form with appended control and error detecting bits. The speed of transmission is dependent upon an external clock pulse, but band rates of 40 K can be achieved.

Behind-the-tape-reader unit

To increase the reliability of transmission in the hostile environment of the workshop, differential line receivers are used; these give high noise immunity and low cross talk. The series-to-parallel receiver takes bits from the line receiver and verifies the information and outputs, in parallel form to the buffer store. This unit has its own external clock, independent of the transmitter, the data being received uses the technique of centre sampling. Also, in order to allow information to be passed intermittently to the machine, a code delete generator was incorporated into the interface, so that when the machine requests information and no information is available, a code delete pulse is output to the machine tool. This is to ensure no zero pulses are transmitted to the machine.

The voltage level and pulse shaping circuit is particular to the system being interfaced and needs:

(i) a zero logic system;

(ii) voltage level of 18 V;

(iii) mark space ratio the same as the Tally Reader.

Software developments of the DNC cell

Normal computer facilities are available such, as;

(i) listing of part-programmes held on disc;

(ii) provision to write statements direct to disc from basic Fortran statements, thus giving computer aided programming and the ability to write family programmes for similarly shaped components;

(iii) management functions of time per part, down time, lead time, etc.;

(iv) error messages output on the main course to assist the computer operator with his various activities and to warn of malfunctions of the system. For example, warnings are given to indicate when the disc is full or when there are errors in the transmission system.

Four main control programmes in assembler language are written, each being called down from the disc into the core memory of the computer when required.

(i) *Programme one.* Information is typed into the computer using the teletype and this information is outputted to the specific machine tool a block at a time using the character command of carriage return.

(ii) *Programme two.* This reads previously prepared tapes in EIA code using the computer tape reader, and converts the code from EIA to ASC II (computer code) and stores the information on disc. As the character is being converted it is also checked for validity, and if it is not recognized, a marker is outputted. The same programme will also output listing of the part programme to the teletype in the normal format.

(iii) *Programme three.* This takes information from the disc for checking and outputs it to the device selected, in the correct format. Provision is made so that the programme can be started at any block and will, if required, be printed out on the teletype as transmission proceeds.

(iv) *Programme four.* To allow for direct output using a high level language like Fortran while calculations are taking place, a programme has been written.

The programme takes characters from the Fortran output and checks the information, adding leading zeros and impressing decimal points as required by the appropriate hard-wired controller. This programme allows for direct circular interpolation and also for the production of family components.

As a result of this work, alternatives are now available such as:

(a) development of other two dimensional standard routines, such as parabola, ellipse and spiral routines;

(b) development of three dimensional standard routines, such as sphere, paraboloid and hyperboloid routines;

(c) development of special purpose routines for practical applications, such as gear and cam routines.

Standard routines

In order to complete the sub-programme system, several routines were written to carry out standard programming operations. The programmes (Milner and Ng 1979) available in sub-programme form are:

(i) a head routine, programme HEAD;

(ii) a tail routine, programme TAIL 1;

(iii) a rapid motion routine, programme RAPID;

(iv) a cutter down feed or drilling routine, programme INFEED 1:

(v) a cutter retreat routine, programme RETRT;

(vi) two write routines, programme WRITE 1 and WRITE 2;

(vii) a straight line routine, programme STLINE 1;

(viii) a circular arc routine, programme NORMAL;

(ix) a cornering routine, programme CORNER 1;

(x) a circle routine, programme CIRCLE.

Main programme

The main programme integrates the required subroutines to form a complete complex routine for a particular contouring requirement, there being no restriction on how many sub-routines are used at one time. Figure 5 shows a typical component produced using the above system of programming routines.

Interactive programming

This is a communicating type of programming where the programmer answers a series of questions posed by the computer, and on receiving the answers, the computer then compiles a machine programme which it stores on disc ready for output to the machine when required.

The computer programme is in two main parts: a Fortran programme, which communicates with the programmer via the teletype, collecting the required data by a series of programmed questions, and an assembler programme, which takes the data passed to it, creates a file under the specified

Figure 5.

name and organizes the data into a form suitable for the machine logic. The machine programme produced on disc can be modified if required, using the normal computer editing facilities.

Three-dimensional contours

To machine a three-dimensional contour, the movement of the cutter tip in a three-dimensional cartesian space is from one point to another, the difference between this and two-dimensional contouring being that the successive points of the cutter may not lie on the same plane.

There are two main ways of computing the position of the required points in a three-dimensional cartesian space:

(i) simultaneously calculating the X, Y and Z coordinates for successive points in each plane such that the cutter moves along the diagonal. The spiral Fig. 6 is an example of a contour based on this routine;

(ii) keeping one coordinate constant and calculating the two-dimensional coordinates in one plane and then moving to the next adjacent plane. The paraboloid Fig. 7 is an example of the three-dimensional component shape which could be more conveniently programmed using this method.

As the three-dimensional equations are often greater than first order, except for plane surfaces, the selection of the appropriate roots of the equation is essential. Direct application of the three-dimensional equations makes possible the development of even more versatile routines, such as the quadratic routine, with which a ' family ' of surfaces instead of only one particular surface can be generated.

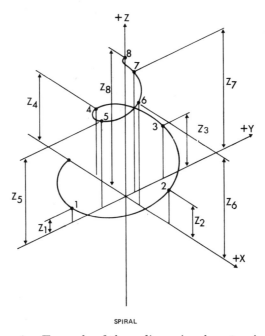

SPIRAL

Figure 6.   Example of three-dimensional contouring.

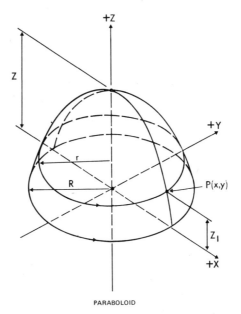

PARABOLOID

Figure 7.   Example of three-dimensional contouring.

Special purpose routines

One of the practical applications considered here is that of cutting Novikov gears. Involute tooth gears which utilize line contact meshing are subject to shortcomings such as limited load capacity, and it has been suggested (Allan 1964–65) that Novikov gears are not subject to this limitation. It can be appreciated that to cut Novikov gears on conventional gear cutting machines would require special form cutters. There are three possible forms of meshing with Novikov gears:

(*a*) an All-Addendum pinion with an All-Dedendum wheel;

(*b*) an All-Dendendum pinion with an All-Addendum wheel;

(*c*) a pair of Addendum–Dedendum pinion and wheel.

The most frequently used combination, because of its higher loading capacity, is the All-Addendum pinion with All-Dedendum wheel meshing, and this standard design is shown in Fig. 8.

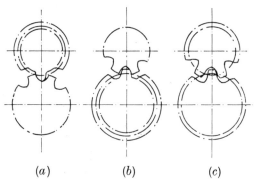

(*a*) (*b*) (*c*)

Figure 8.   Novikov gears.

Salient ideas of the gear routines

Basically all the contours are circular arcs and it is possible to define all the points which marked the change of contour on the tooth profile, With circular arcs, cutter compensation is readily achieved by a radial shift of an amount equal to the cutter radius value. From a practical point of view, the functional contours are the convex and the concave profiles. The less important contours such as the flank radius, tooth tip, etc., are to be modified for convenience. In this case, both the flank radius and the radius blending the intersecting concave profiles have been fixed to be equal to the cutter radius, thereby reducing the number of points to be defined and the amount of computation.

As the contour of each tooth on the same gear is identical, only points within one tooth pitch need be defined. In addition to the points on the contours, the various centres of the arcs have to be located as well, requiring the use of geometrical equations in polar coordinate form.

A careful examination of the various parameters used in the standard design revealed that it is possible to define the gear precisely by only a few basic parameters, namely, module, number of teeth, pressure angle and the gear centre coordinates. After reading in the data, the computer software goes through a series of design parameter calculations, followed by the usual commands for rapid motion, drilling and feedrate setting. Then comes the section to calculate the geometrical parameters for the respective paths, the completed component being as shown in the photograph.

Conclusions

It has been demonstrated that once the necessary interface has been built between the machine tool and minicomputer, the range of facilities can be quite enhanced.

On the software side the following conclusions may be drawn:

(1) In designing the software routines for linear interpolation, it has been found that the non-incremental method using chords is the most promising method. Using this method, the respective contouring routines developed are free from cumulative errors. Its most important asset, however, is that it enables the direct application of the equations of the contours themselves. This means that any geometrically defined contour can be programmed and generated using this method without difficulty. In cases where contours cannot be geometrically defined by equations, it is also possible to break the contour down into sections, define each section by suitable points and approximate the contour between these points with the routines developed using this method.

(2) From the results of the tests on the circle routine, it is deduced that the contour accuracy which the developed routine can generate is only subjected to limitations in the capacity of the component and that of the machine used. For the machine used, there always exists an accuracy limit which marks the best practical accuracy the machine can reliably attain. As a result, to specify more accurate tolerance value for the routine used would not produce a more accurate contour from the machine. For the present machine, the accuracy limit and the corresponding tolerance for contouring were both found to be approximately 0·001750 in. The most probable cause for the machine errors is the steady state lag in the drive mechanisms. Neither the cutter dwell due to the acceleration–deceleration cycle of the machine control system during change of path directions nor the possible cutter deflection during machining contribute significantly to the limiting error of the machine.

(3) The successful implementation of the Novikov Gear Routines not only enhanced the correctness of the working principles involved, but also sees the tremendous potential of the developed standard routines for extension into practical applications. There are many advantages to be gained by using these special purpose routines as compared to conventional methods. Taking the spur gear as an example, to machine gears of different tooth sizes, the number of cutters required will be as many as the number of tooth sizes themselves. As these are form cutters, the cost of both manufacturing and stocking them can be quite high. Apart from that, if, say, for optimum performance some other parameters, such as the pressure angle, are also required to be varied, then for every tooth size there can be more than one shape proportion. The consequence is that an even greater number of cutters is required. Using the special purpose routine, any desired variation can

equally be achieved simply by changing the input data for the respective input parameters, and one can vary as many parameters as desired. Instead of many form cutters, only a small range of standard cutters is necessary. Another important point here is that by using the special purpose routine, we can replace mechanical indexing of the gears with mathematical indexing. This not only saves the cost of using the mechanical indexing equipment but also eliminates its source of errors. All these would add up to significant cost savings without loss of product quality. Other advantages of the special purpose routines are similar to those of the standard routines. The Novikov Gear Routines are but one example of how the standard routines can be extended into practical applications. There are many other areas where they can be very useful in cost savings.

(4) From the present form of the developed software routines, to achieve contouring capability for an NC machine without such facilities, it is not necessary to use very complicated special purpose language for part programming, such as APT, which not only requires a large storage but also a special compiler. Ordinary Fortran or even Assembler Language can serve the purpose just as well. Proper design of the input data format has made possible the necessity for only very few basic data to generate a large quantity of output for contouring.

(5) The development of the paraboloid routine shows that the principles involved in developing two dimensional routines can be extended quite easily for three-dimensional contouring purposes. This is because, as in two-dimensional contouring, the cutter in three-dimensional contouring also moves from one point to another. By generating all points in a plane parallel to one axis before generating points in the next adjacent planes, two-dimensional routines can be used to create three-dimensional surfaces.

(6) The conversion of the existing software routines into a system of subroutines sees the establishment of a more versatile system of software interpolation. The modular style of the subroutines makes their manipulation much easier, and occupation of the computer storage space is greatly reduced. It also becomes possible to develop from them any desired complex routines, and expansion of the whole system is easy.

(7) The successful working of the software interpolation system as a whole proves convincingly that conversion of a non-contouring NC machine to one which can perform contouring can be accomplished quite easily in software. The chief advantage of software interpolation as compared to hardware interpolation is that, owing to its inherited adaptability to changes, it can always be modified to suit new working environments without the danger of becoming obsolete. Other advantages of interest include its ease of manipulation and versatility. For instance, the input data format can be altered to suit a particular use, cutter compensation can be achieved easily, extension for complicated contouring application is relatively convenient and both two-dimensional and three-dimensional contouring are possible. As for the cost of acquisition and maintenance, software interpolation also has greater appeal. Where future trends are concerned, there is an increasing degree of computer involvement in industrial automation of process control, and a

software interpolation system can be easily integrated with other aspects of control, such as inspection, work-handling and maintenance, to form a wider and more flexible system.

Illustre les différentes étapes de développement requises pour le montage en rattrapage d'une machine à CN existante, en donnant les détails appropriés sur la liaison de l'interface. Sont également signalés certaines développements du logiciel permettant d'élargir les performances de la machine-outil lorsqu'elle est connectée en ligne à un mini-ordinateur. La production d'engrenages Novikov est mentionée.

Es werden die verschiedenen Entwicklungsstufen, die für den Wiedereinbau einer vorhandenen NC-Maschine erforderlich sind, beschrieben, wobei die entsprechenden Einzelheiten über die Interface-Verbindung angegeben werden. Es wird auf einige Software-Entwicklungen hingewiessen, welche den Bereich der Werkzeugmaschine erweitern, wenn diese laufend an einen Minicomputer angeschlossen ist. Außerdem wird die Herstellung von Novikov-Getrieben erwähnt.

REFERENCES

ALLAN, T., 1964–65, Some aspects of the design and performance of Wildhaber–Novikov gearing, *Proc. Inst. mech. Engrs*, **1–3,** 931.

CROSSLEY, T. R., and LEWIS, W. D., 1974, *DNC Prototype, Conference on Direct Numerical Control*, Loughborough University of Technology, pp. N1–N20.

INABA, K., 1973, The present and future in development of machine tool system with group automatic control, *Dig. Jap. Ind.*, **66,** 15.

MILNER, D. A., 1974, Adaptive control of feedrate in the milling process, *Int. J. mach. Tools*, **14,** 187.

MILNER, D. A., and NG, K. H., 1979, Software interpolation for on-line computer control (to be published).

Reprinted from Manufacturing Engineering, January 1977

1. PALLET SHUTTLE SYSTEM includes shuttle car with pallet changer, and load/unload stations.

Increasing Machine Utilization In a DNC Manufacturing Line

Here's a solution to the material-handling problem in DNC – a pallet shuttle car under control of the central computer

ROBERT N. STAUFFER
Associate Editor

DNC MACHINING SYSTEMS are now well established as an approach to improving productivity and reducing manufacturing costs. Along with an increasing acceptance of the concept, there is also a continuing effort to refine these systems to meet changing manufacturing requirements. Sundstrand Machine Tool, Belvidere, IL, is a case in point. The firm recently introduced a new pallet shuttle system featuring a computer-controlled shuttle car designed to run between a number of NC machines. Parts are transferred from a pallet waiting station and loaded/unloaded at the machines in a programmed sequence that provides maximum machining efficiency.

The Omniline. The versatile shuttle car is one element in the Sundstrand Omniline, a computerized fully automatic system for batch manufacture of a variety of complex parts. In these lines, all of the machines, tools, fixtures, controls, and workhandling devices are selected and integrated to achieve maximum production efficiency and economy. A typical line includes one or more Omnimil NC machining centers, vertical spindle NC turning and boring machines, multiple spindle drilling machines, and any other machines that a customer may already have in-house. With this modular arrangement, it's also possible to start with one machine and a load/unload station and tie in additional machine tools as needed in the future. All units are controlled through a DNC system,

with a SWINC soft wired control at each machine.

The pallet shuttle system, *Figure 1,* provides automatic machine loading/unloading and parts transfer through all of the processing steps. Major elements of the system include the shuttle car, standard pallet changer, and load/unload stations. Sundstrand points to the system as providing all the benefits of the transfer line concept, but with the further advantage of flexibility to handle dissimilar parts.

Shuttle Car. This powered transporter is mounted on four machined steel wheels and rolls on a set of steel rails connecting the various machines and load/unload stations. The car is driven by a d-c motor and timing belt, driving through a pinion gear running against a gear rack mounted on the side of one of the rails. During acceleration and deceleration the positive-type drive eliminates slipping due to an oily track or varying pallet load weights. Maximum velocity is 300 fpm (1524 mm/sec). Braking is done with a disc brake on the drive train shaft. The car handles either 32″ (813 mm) or 42″ (1067 mm) diameter pallets, each with a capacity of 8000 lb (3600 kg).

Automatic control is in Sundstrand's DNC Omnicontrol mode, or in a CNC stand alone mode with SWINC control. Electric, hydraulic, and control components required for car operation are carried on board. An operator control panel is provided for maintenance and setup purposes. There is enough room on the car to accommodate two people if required.

Control signals and electric power for the car's drive motor are supplied from an overhead, self-standing bus bar system. The elevated bus bar is out of the way of machining oil, chips, and other debris for safety and reliability. Additional safety features include an extended bumper at each end to detect any obstruction on the track and signal for an automatic emergency stop. The car stops within 1 1/2 feet (457 mm), which is before the bumper fully retracts into the car's structure. For further protection, a latching mechanism prevents the pallet changer shuttle forks from extending unless the car is parked at a machine or load/ unload station. A fixed cam at the station opens the latch mechanism to permit the forks to extend.

At each machine and load/unload station, the car is held in position by an electronic servo system rather than mechanically with shot pins or other means. An encoder determines the car's exact longitudinal location within ±0.015″ (0.38 mm). One advantage of the servo system is that it permits easy adjustment of car position to compensate for variations in machine location.

Universal Pallet Changer. The automatic pallet changer is basically the same standard changer offered with the Series 80 Omnimil. In a typical load/unload cycle, *Figure* 2, the shuttle car stops at the machine, and the changer rotates to the "load" position. The machine table raises pallet A to be unloaded, the hydraulically driven pallet changer forks extend under the pallet, and it is then lowered onto the forks. After retraction from the machine, the shuttle forks rotate 180°, bringing pallet B into position ready for loading. The forks then extend to the machine where pallet B is picked up by the table. As the forks retract with pallet A, the machine table lowers and clamps pallet B in position ready for machining. The pallet changer shuttle then rotates 90° to the "transport" position ready to carry pallet A back to a load/unload station or to another machine. Total time for a pallet change is approximtely 30 seconds.

Precise pallet location is achieved with a Curvic coupling. One-half of the coupling is on the machine table, the other half on the bottom of the pallet. Pallet location is maintained within 0.0005″ (0.013 mm) both longitudinally and vertically, and it is hydraulically clamped at 2000 psi (13 790 kPa). A

2. PALLET CHANGER in operation. Above: changer arrives at machine in "transport" position; Right: forks extend to retrieve first pallet/part unit; Below: changer has rotated 180° ready for loading second part/pallet.

protective cover slides over the lower half of the coupling after the pallet is lifted off.

The load/unload stations can be operated manually or under automatic control. Also, they can be either the stationary type, or designed for manual pallet rotation to provide easy access to the fixture. Operator instructions and system status are displayed on an information panel.

Sundstrand's new pallet shuttle system is expected to further enhance Omniline's ability to reduce direct labor costs and part setup time at the machines, and to increase machine utilization and system flexibility. ∎

Presented at CAM '78

Computer Aided Information Flow in Small Batch Production

By A.J. McTernan and J.L. Murray

SUMMARY

The small batch production associated with high technology products inevitably carries the penalty of high design overhead costs. A computerised design system is discussed in which the data produced can form the basis of an effective information flow system covering several management functions, including estimating, anticipated work load, production and stock control. Drawings, manufacturing data, and numerical control tapes can be automatically produced where appropriate.

INTRODUCTION

Computer methods are well established in some areas of engineering design analysis. The integration of the design process with manufacturing and management functions can also be achieved using computer-based systems. One of the most fruitful areas for this approach is one where there is a family similarity between products in the range, without necessarily too close an identity between products. This similarity enables a limited catalogue of design procedures to cover a very wide range of possible product configurations. The design procedures can be used to trigger modules which generate the associated estimating, manufacturing and management information.

The expansion of computer aided design techniques in industry has not been as fast as was predicted. This is mainly due to the fear of high development costs in the computer system and lack of flexibility in the completed schemes. No doubt this is in part due to the scarcity of staff who combine engineering expertise with an appropriate knowledge of computer-based information systems design. In this context, Hatvany[1] has suggested greater use of modularity in the system in order to minimise development costs and facilitate technology transfer.

The position of the designer as a component in the design loop, taken for granted in conventional design techniques has tended to be neglected in a number of computer aided systems. Warman[2] argues that the designer should participate in the design loop as a problem solver.

An integrated approach is advocated by the authors in which the engineering designer is placed firmly in the design loop, and the entire system is based on a modular structure representing self-consistent stages. The prime benefit of such an approach is that it gives the manufacturer speed of response in a competitive quotation /

quotation situation, without losing the flexibility available
using manual methods of designing to a customer specification.
Also, in the integrated system, if an order is placed the problem
of transfer of design information to the subsequent manufacturing
stages is minimised.

This approach is particularly advantageous in a small batch
production environment where the overheads account for an above
average proportion of product cost.

CHARACTERISTICS OF SMALL BATCH DESIGN

Small batch production leads inevitably to relatively high
overhead costs, associated with a number of different organisational
functions. Some of these are detailed below.

High technology products, in particular, require skilled sales
engineers and their work calls for cost estimates to be produced
quickly, yet accurately enough to commit the manufacturing company
if the quotation is successful. Accurate estimation requires
identification of the product sub-assemblies, their design to
a depth sufficient to prove their physical compatability and
performance, and the evaluation of costs associated with production.
The greater the depth to which the design and detail estimating
is taken, the greater the confidence one can have in the resulting
estimate. On the other hand, the greater the depth, the greater
the cost associated with this stage. While it is true that if
the tender is successful, the preliminary design and costing
work forms a useful base for subsequent detail design, many
industries must expect a low success rate for tenders. For such
industries, in particular, it is important to try to improve the
efficiency of the tendering procedure while maintaining the speed
of response.

When an order is received, the preliminary design information must
be confirmed before the labour-intensive stages of detail design
and drafting are undertaken.

Information from the detail design forms the basis of production
planning, material requisition and tooling requirements, leading
to the eventual manufacture of the product.

The nature of small batch production means that the costs
associated with the generation and transfer of the information
relating to the product have to be carried by the small number
being produced.

PRODUCT FAMILIES

There is a universal desire to standardise components and
materials, in order to reduce design, manufacturing and stock
holding costs. It is possible to go some way towards variety
reduction by using family properties applied across a range of
products. Frequently a wide range of products is manufactured
from a selection of similar components.

The similarity may be one of shape, method of manufacturing or
physical environment.

While similar components can be built up into similar assemblies it is necessary to consider the problem of dimensional and functional interfacing between sub-assemblies.

One approach to the problem is to create families of similar parts which, in a given product design, can be scaled to suit the particular application. This may involve the geometrical properties, the material properties or the functional requirements of the product e.g. strength, stiffness, weight.

It is important in this context to appreciate that any part can be completely defined by its material and a geometric description. In some cases the complete material definition may be complex, e.g. if heat-treatment or surface preparation is necessary. The geometric description may consist of a shape code together with dimensional parameters. In this way a large variety of component shapes can be specified concisely.

Physical and other properties of any component so described can be determined by standard analytical methods, e.g. weight, cost, vibration characteristics.

Examples of suitable product ranges exist in the design and manufacture of cranes, structural steelwork, pipework systems, pressure vessels and heat exchangers.

Brief consideration of a heat exchanger product range, for example, shows that (as in Fig.1) there will be two covers, shell, a number of flanges leading pipes into the pressure vessel and tubes within the vessel, supported by tube plates. The thermal designer could state a need for a specific number of tubes of specified dimensions, operating under specific environmental conditions of temperature and pressure. The task of the mechanical designer is to select a structure which is physically compatible with the external system, meets the requirements of the thermal designer and satisfies the appropriate mechanical code of practice as regards strength, etc.

INFORMATION FLOW IN SMALL BATCH PRODUCTION

Fig. (2) shows the flow of information in a simplified scheme representative of a system for the design and manufacture of small batches of high technology products which have a family similarity and the design of which follows a well established procedure. This consists of two phases.

The first phase centres round the preliminary system design leading to a customer quotation. It is generally triggered by a sales enquiry, following which the designer creates a design scheme consistent with the company product range. With the aid of proven design methods, the feasibility of the concept is assessed. If analysis of the performance reveals that it does not meet the design specification or the designer, from experience, considers optimising changes advantageous, a further iteration of the design process may be undertaken. The policy of using standard materials, components and tooling will of course influence the design decisions.

The output from this preliminary design study forms the basis from which the quotation and installation details can be produced.

The second phase only occurs if the quotation is successful and an order is received. The preliminary design information is revised to take account of any minor modifications and the detailed design is completed. Another triggering takes place at this point. Information from the detail design passes to purchasing and stock control, production planning and control and allows the compilation of other management statistics. It is possible to use the design information to generate numerical control tapes where applicable.

COMPUTER ASSISTANCE

To anyone with experience of attempting to enforce company standards it comes as a nice revelation that a computer based design system will only follow the rule book.

The computational speed of the computer can be exploited to permit the use of more sophisticated design procedures, and consequently more accurate design calculations. This increased degree of confidence in the analytical results enables designs to be made within closer limits. In addition the design engineer can investigate various design configurations to within the limited time scale available during the tendering period. Also, in a competitive situation, the ability to make accurate estimates of cost allow the determination of tender prices with greater confidence.

If the design procedures used in the preliminary stage are no less precise than those necessary in the detail design stage the estimating information is made more accurate and the transition to detailing eased considerably.

SYSTEM DESIGN

Referring again to Fig. (2) the design procedures will be implemented in a computer based system by a library of design algorithms, to cover the range of different types of components and sub-assemblies. The algorithms will call upon files of standard components, e.g. fixings and fastenings, materials and available tooling under the control of the designer. The system will at each stage check the design for geometrical and functional compatibility. Any discrepancy can be signalled to the designer for his decision either to modify the design or to accept a default modification generated within the system. For example in pipework system design it may be found necessary to increase the specified thickness of a pipe in order to satisfy a particular code of practice.

The output from the preliminary design phase is a material and geometry file. From this file estimating details can be produced using appropriate costing algorithms covering both component manufacture and assembly. Quotation drawings can also be directly generated using a digital plotter. This is particularly the case where standard assembly ranges exist.

The material and geometry file also serves as the information link to the post-order operations.

Once an order has been confirmed the detailed manufacturing documentation must be produced. In the system advocated, the vast majority of the detailed design has been carried out during the preliminary system design. The generation of detail drawings, manufacturing templates and associated data is relatively simple for families of components. A method successfully applied by the authors is based on the use of a geometry generating algorithm for each component family. The actual dimensions for a particular component are controlled by parameters input from the system geometry file. Individual geometry generating algorithms for complex shapes are created by linking a series of simple shape generators. Assembly drawings for standard ranges can be produced in the same way.

Where numerically controlled machine tools are employed as part of the production process, automated control tape preparation is again straight-forward. In this way, by directly accessing the geometry file, potential transcription errors are avoided. The dimensions and other critical characteristics of each component would be checked for compatibility with the proposed manufacturing process before tape production is undertaken.

The material requirement must be known at an early stage in the production cycle. From the full quotation information already produced material details are available for forward planning. Therefore no delay in issuing material requisitions need be experienced.

Individual components in the same family will normally be produced by a standard combination of processes depending for example upon the size. This standardisation allows proceduring algorithms to be prepared. The output from these will list the individual operations together with allowed times, machine and tooling requirements, etc. Such planning data, fully detailed, and to a specific format can readily form input to the commitment file of a computer-based production control system.

A further by-product of this systematic approach is the ability to produce a wide range of management statistics. For example from the files of sales enquiries and quotation information, sales forecasting can be undertaken. These forecasts can be expressed in terms of sales value and of anticipated shop loadings.

PROGRAMMING TECHNIQUES

Once the overall information flow network has been established a computer program suite can be developed on a modular basis. Even if a fully computer-based system is not to be implemented immediately the interfaces between modules must be fully specified.

Each unit in the information flow network can be individually 'computerised' enabling the computer system to be developed naturally as experience and confidence are gained.

The authors have developed a major part of an integrated information flow system for an engineering firm. The program suite, written throughout in a high level programming language, was developed using structured programming techniques. This resulted in the production of a reliable, portable and easily maintainable package at minimum cost. The package was developed so as to run in batch mode and in interactive mode. One advantage of the interactive mode is that sales engineers can take a portable terminal to a prospective customer and thus the customer can be involved in the design loop,

DISCUSSION

The experience gained by the authors in implementing such a system has established that the approach is a feasible one. By developing the program modules at a controlled rate, the user management gains familiarity and confidence in each progressive phase of the system. The success of one phase motivates their efforts tc complete the next. On the other hand the change over from manual to computer based processing does not happen too quickly for the personnel involved.

As a result of the system implemented it is estimated that there was a 40% reduction in the staff required to maintain a prompt, reliable design quotation service. The reduction in time scale permits the relatively easy possibility of a second iteration of the design cycle. In order to keep the information base of the computer system in line with the cost-time relationships associated with current production techniques, the company management have set up a scheme to provide an updating feedback path from actual works costs to the empirical relationships which form the basis of many of the costing algorithms.

The well-structured nature of the programs has allowed additional components to be incorporated in the repertoire, and modifications to the design procedures of existing components to be made easily.

ACKNOWLEDGEMENT

The authors would like to thank Professors A. Balfour and T.D. Patten for encouragement and the provision of facilities.

REFERENCES

1. Hatvany, J.,   "The Use of CAD/CAM Systems in Manufacture", IIASA Research Memorandum RM-74-22: Laxenburg, 1974.

2. Warman, E.A.,   "Man's rightful place in the design loop", Data Systems, August 1975.

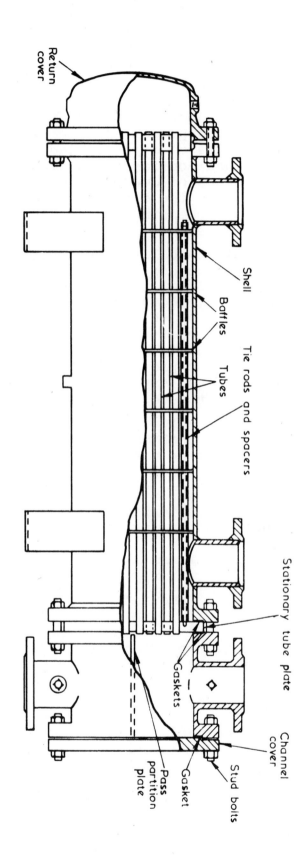

FIG. I. TYPICAL HEAT EXCHANGER LAYOUT.

Return cover

Shell

Baffles

Tie rods and spacers

Tubes

Stationary tube plate

Channel cover

Gaskets

Gasket

Stud bolts

Pass partition plate

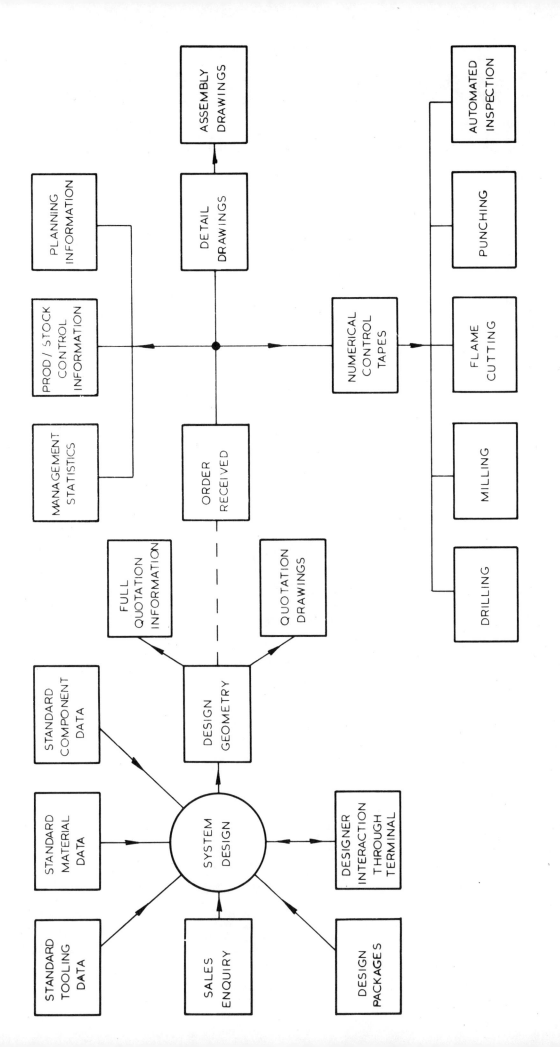

FIG. 2   INFORMATION   FLOW   SYSTEM

Reprinted by the Society of Manufacturing Engineers from IRON AGE, November 20, 1978; Chilton Company; 1978

TAKING THE WRAPS OFF FLEXIBILITY IN MANUFACTURING

By Raymond J. Larsen

ABMSs hold the promise for unclogging some of the industry's severest production bottlenecks in the 1980s and beyond.

Automated Batch Manufacturing Systems (ABMSs), which promise to unclog some of industry's severest production bottlenecks, are beginning to find widespread use in United States manufacturing, setting the stage for the biggest shootout U.S. machine tool builders have seen since the introduction of numerical control (NC) technology 30 years ago.

Cincinnati Milacron, Inc., Kearney & Trecker Corp., White-Sundstrand Machine Tool, Inc., Giddings & Lewis, Inc., Ex-Cell-O Corp., Bendix Machine Tool Corp., Ingersoll Milling Machine Co., Snyder Corp., Cross Co., Kingsbury Machine Tool Corp. and Yamazaki Machinery Corp., U.S. subsidiary of Yamazaki Machinery

Works Ltd. of Japan, are just some of the companies vying for a piece of the rapidly growing ABMS market. All are spending millions of research and development dollars these days to produce ABMS hardware and control systems in the belief that manufacturing systems will account for a major portion of machine tool revenues in the 1980s and beyond.

No one can say for sure just how large the ABMS market is at present. An informal poll taken recently by *Iron Age* showed that manufacturers have placed orders for approximately 35 systems to date, but companies that spend up to $20 million a shot for an ABMS usually don't want their competitors to know what they're up

to, and the number of systems now in the field or on order could conceivably run as high as 40.

Whatever the figure, machine tool industry experts agree that ABMSs are the fastest growing segment of the market, and say sales could easily tip the $200 million mark in five years.

"The growth we've seen in systems orders has been in excess of 50 pct a year during the past several years," an official of one of the nation's biggest ABMS builders said during a recent interview. "In five years the annual market could easily reach $150 to $200 million. The systems approach has moved out of the concept stage and into the business stage."

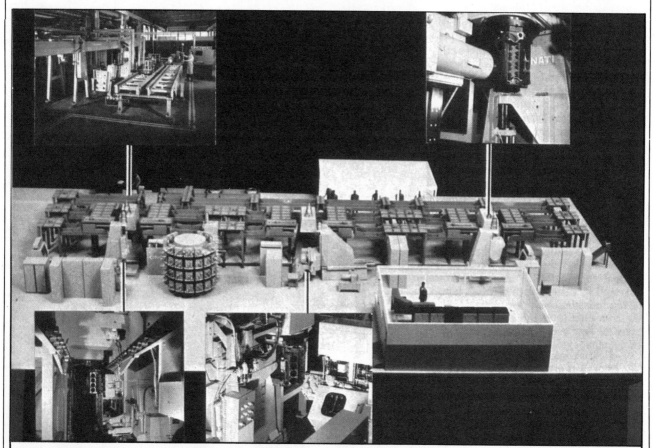

Cincinnati Milacron built this experimental ABMS several years ago. Although it was never sold it's still one of the most flexible, complex and innovative systems ever built.

"A great deal of pressure is building in the systems market," added a top-ranking official at a competing company. "The major constraint right now on construction of systems is the fact that most machine tool builders are swamped with orders for standard machinery, they just can't put more work out the door.

"Potential buyers of systems are in the same boat. They haven't got the resources to develop and implement systems. They're too busy meeting current demands. The pressure continues to build, however, and we feel there will be a tremendous burst of orders in the next several years. It is not unreasonable to believe that the U.S. market will reach $125 to $150 million in five years. That's 10 to 15 systems, and that's not unreasonable at all."

Just what is an ABMS and what can it do for manufacturing?

Even the National Machine Tool Builders' Association (NMTBA), and the special ad hoc committee it formed last year to delve into the question, has a hard time getting a handle on it. But an ABMS can generally be defined as an electronically controlled assemblage of flexible and dedicated machine tools linked by material handling equipment in such a way as to convert stop-and-go batch manufacturing into continuous or nearly continuous processing.

In the current state-of-the-art, ABMSs fall into two categories, those that deal with rotating parts that primarily require turning-type operations, and those that deal with prismatic parts of the type generally handled on milling machines and machining centers.

These systems can be further broken down into two types: Those that can handle families of parts in any order, known as random or flexible systems, and those that process only one part, or parts with only small differences, in sequence, and known as sequential or dedicated systems.

Random systems require direct numerical control (DNC) structures to accomplish their complex manufacturing missions and are considered the creme de la creme of the ABMS market. Few companies can build a big random system.

Dedicated systems employ CNC, NC, or even programmable controller (PC) structures to do their work. They're not as fancy and are generally less expensive than random types, but the competition amongst builders for shares in this end of the market may be the stiffest.

Regardless of whether it is a random or dedicated type, the point of the ABMS is always the same: To cut unproductive manufacturing time to a minimum, thereby increasing—often dramatically—the amount of time expensive machine tools actually spend making chips and moving components through the production cycle.

Machining centers are a key element to ABMS. These big, flexible machines, which can store 90 or more tools, including small, multiple-spindle tool heads, can perform a wide range of cutting operations on

---

"Systems are going to be a big part of metalworking in the future."

Thomas Shifo, general sales manager, White-Sundstrand

---

many different workpieces. Linking several machining centers together with automated material handling equipment can eliminate a lot of conventional downtime attributable to the constant refixturing and handling inherent in stand-alone machine tool operation.

Head indexers and changers are another key element to the ABMS. These three-, four-, and five-axis units are really mini-systems in themselves. The typical head changer consists of a programmable drive unit and a storage rack or carousel holding up to 90 dedicated, multiple-spindle tool heads. Electronic or hy-

draulic mechanisms move the heads into the drive station, either randomly or in sequence, and the head is then brought to bear on the workpiece. Most machine tool builders competing for new business in the ABMS market have developed some type of head indexer or head changer, and several of these builders have sold head indexers and changers as stand-along machines.

The machining centers, head changers, and other machining elements in the typical ABMS almost always retain their own NC or cnc control unit. This gives the system a maximum amount of flexibility and, in the case of DNC systems, permits continued operation of the system in the event of a malfunction of the central computer.

ABMSs aren't for everybody.

Manufacturers who produce a lot of parts in small numbers (say 200 different types in batches of 10 or 20) can't justify the high cost of ABMSs compared to stand-alone nc machinery or well-designed Group Technology-type manufacturing cells.

Similarly, producers of a few parts in large numbers (say one or two types in volumes of 10,000 or more) will not be able to squeeze enough production out of an ABMS, even a dedicated ABMS.

But many manufacturers whose production requirements fall somewhere in between will find that ABMSs can truly work wonders for them.

Under ideal conditions, a well-designed, well-implemented ABMS costing $5 million can pay for itself in one

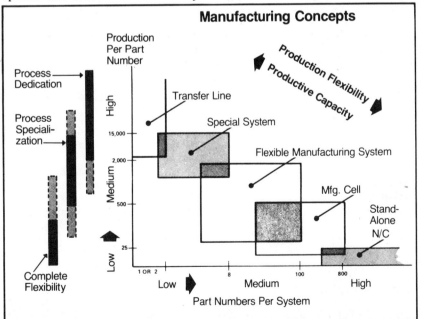

Manufacturing Concepts

This chart represents Kearney and Trecker's philosophy regarding component manufacture. The role of the ABMS is clearly outlined.

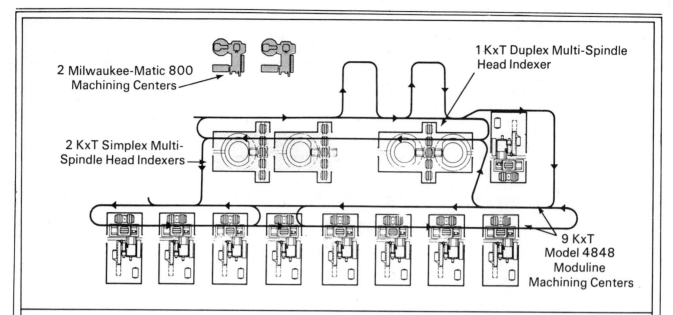

2 Milwaukee-Matic 800 Machining Centers

1 KxT Duplex Multi-Spindle Head Indexer

2 KxT Simplex Multi-Spindle Head Indexers

9 KxT Model 4848 Moduline Machining Centers

Diagram shows the ABMS built for Avco Corp.'s Avco Lycoming division in Williamsport, Pa., by Kearney and Trecker Corp. The system machines aluminum crankcase halves for Lycoming aircraft engines.

year and increase production of difficult workpieces by several hundred per cent. That's why more and more manufacturers are plunking down their hard-earned cash for systems these days, and why machine tool builders are gearing up for the biggest battle since the shootout at the OK Corral.

Interestingly, Cincinnati Milacron—which has dominated U.S. machine tool building for several decades—finds itself in the unfamiliar role of underdog in the battle for dominance in the ABMS market. For clearly, Milwaukee-based Kearney & Trecker has taken the lead in the ABMS market these days, both in number of systems sold (both random and dedicated), and in number of system components developed and tested. And the upcoming merger with Cross—another machine tool builder active in the ABMS market—promises to make the proposed Cross & Trecker operation even more formidable in the years to come.

K & T got into the ABMS competition early, and the $5 million system it installed to machine tractor castings at Allis-Chalmers Corp.'s West Allis, Wisc. plant in the early 1970s was one of the first big random systems seen in U.S. industry.

Since then the company has shipped or taken orders for four additional random systems. They include a system for machining truck axle components at Rockwell International Corp.'s axle plant in Newark, O., an expanding system to machine airplane engine crankcases at Avco Corp.'s Lycoming division plant in Williamsport, Pa.; a system to machine XM-1 tank engine components

at Avco's Lycoming division plant in Stratford, Conn., and a system to machine transmission components at a plant K & T declines to name.

K & T also has shipped or taken orders for several dedicated systems, including two systems for a large construction equipment manufacturer; one system for an unnamed manufacturer of heavy duty axle housings; a fourth system for ma-

---

An ABMS can cost $5 million, pay for itself in one year and increase production several hundred pct.

---

chining large frames for an unnamed manufacturer of construction equipment, and two systems scheduled to be shipped to England. One of these is designed to machine large frames for a construction equipment company. The other is designed to machine parts for the automotive aftermarket. The names of the manufacturers have not been revealed.

The sale of that many systems has permitted K & T to develop and prove out a number of innovative ABMS components, including machining centers to 50 HP, Simplex (single-station) and Duplex (dual-station) head indexers, several types of head changers, a special machining unit capable of turning and facing palletized

workpieces, a variety of dedicated machining stations, random and sequential parts loading and handling systems, and the electronic controls needed to keep all of them running. These components give K & T an even bigger edge over its competitors.

Much of K & T's success to date can be traced directly to Paul R. Haas, vice president and general manager of the company's Special Products Division and one of the best ABMS design and marketing minds in the industry today.

Mr. Haas' approach to the market places a great deal of emphasis on longterm manufacturing strategy. He sees the ABMS (Flexible Manufacturing System in K & T parlance) as a hedge against many of the manufacturing pitfalls inherent in the life cycle of a product. As he sees it, the ABMS can serve as an important buffer between limited production and mass manufacture. Thus, an ABMS might be a logical step en route to installation of a high-cost, low-flexibility transfer line in the product's heyday. It might also be a nice way to handle production in the product's twilight years, when manufacture on a transfer line is no longer viable.

Mr. Haas points to the need for replacement parts in the auto industry as an excellent example of how an ABMS can help improve an auto maker's manufacturing efficiency. "A lot of these companies continue to produce replacement engine blocks on transfer lines long after there is no economic justification for that kind of operation," he said.

Mr. Haas also believes that ABMSs work best when they're phased into a plant over a period of time. "Our strategy," he said, "is to build and ship a system on a modular basis. It doesn't make sense on a major system to try to assemble it in our plant, run it in, then break it down and ship it to the customer. We design our systems to permit installation in stages." Mr. Haas noted that K & T also tries to design its systems so that they can be changed over to accommodate different part mixes down the road, or added onto if volume or product mix warrants it.

A recent tour of K & T's main plant gave a good indication of the kind of work the company is doing these days.

Two systems for a construction manufacturer were under construction.

For the first system, designed to machine seven-ton earth mover frames, K & T has developed its biggest machining center yet, a 50-HP traveling column type designated the Maxi Modu-Line and capable of handling tools weighing up to 800 pounds each.

K & T also is using an extremely large head changer in the system. The changer can store up to 30 heads on each of three levels, though that many heads probably will not be needed for the workpiece that the system is designed for.

This basically dedicated-type system was said to be the most costly ever constructed by the company.

Also on the floor was a shuttle-loading, partially random type system designed to machine several different sizes of track guide mechanisms. Especially innovative in this system is the use of a large revolving drum station containing several spindles for rough and finish boring. This unit permits precision boring of a particular pattern on different locations on the workpiece, as is required when the component must be modified to accommodate various options.

Although K & T holds the lead in ABMS production at present, manufacturers would be wise to wait a while before writing Cincinnati Milacron off as an also-ran.

In the early 1960s, long before ABMSs were a gleam in Francis Trecker's eye, Milacron designed and built what may well be the most complex and innovative random ABMS system ever.

The system, built with some assistance from Ford Motor Co., was designed to machine a wide range of dissimilar parts Ford planned to use in a small gas turbine engine it was designing for use in passenger cars. The engine never got off the ground, Ford never bought the system, and

eventually, it was relegated to the scrap heap (Milacron felt the system was basically a prototype that could not be sold on the open market). But before the system was junked, Milacron developed a mountain of data on the care and feeding of big, random manufacturing systems. Some innovations built into the unit, including its through-the-spindle coolant system and its high-speed (two-second chip-to-chip time) tool changer, eventually found their way to Milacron's line of standard machines. Others, such as its massive head changer, became components in Milacron's ABMS arsenal. (Still others, such as its unique upside-down pallet conveyor system, which eased handling of chips from the cast iron parts it was machining, have been shelved. Milacron officials smile when

ABMSs are the fastest growing segment of the market.

the upside-down pallet system comes up. It's far too expensive to be included in current systems, they agree, but it was one hell of an idea.)

Although Milacron has yet to sell its first random system, it has sold a number of dedicated systems. Milacron doesn't like to be pinned down, but it has sold as many as eight (depending on which official you talk to) of these units to U.S. manufacturers, including Caterpillar, Eaton Corp., Westinghouse Electric Corp., and Chrysler Corp.'s Sterling Defense Division in Sterling Heights, Mich. This division holds the Army contract for the justly famous XM-1 battle tank.

As is the case with K & T's Mr. Haas, Charles F. Carter, Jr., manager of advanced machine tool systems for Milacron, is convinced that the ABMS is a concept whose time has come.

"We're seeing a lot of interest from a number of markets," he said recently. "Aircraft engine, agricultural equipment, construction equipment and similar manufacturers are showing a great deal of interest these days. Quotes are up. Last year we quoted one system. This year we've quoted more than three already.

"The fact of the matter is that the technology is there. Now, it's up to the progressive manufacturers—the Caterpillar Tractor Co.'s of the world—to come forward."

Frank Curtin, vice president for machine tools in North America, noted that Milacron's overall strategy is to sell all segments of the systems market.

"Of course we're interested in selling as many big systems as possible," he said, "but systems aren't all of the big Caterpillar type. We're just as interested in selling small systems such as the robot manufacturing cell we exhibited at the International Machine Tool Show in Chicago.

"There is no clear leader in the systems market at present, and we feel that our broad line of machine tools, coupled with our capabilities in softwear, which are second to none, put us in an excellent position in the systems market."

As at K & T, a tour of Milacron's Oakley production facilities was extremely informative.

Milacron currently is building three dedicated systems, one for the machining of large frames for Caterpillar, one for the machining of XM-1 tank hulls, and one for the machining of XM-1 torsion bar housings.

Nothing can adequately prepare the viewer for his first look at the XM-1 tank hull system. It is gigantic and highly innovative; it is a brilliant piece of machine tool systems engineering, flawlessly executed.

For one thing, the hull, fabricated of specially formulated armor plate, is absolutely massive, tipping the scales at around 15 tons. To move it through the system, Milacron had to devise a complex water-cushioning pallet transport system, the only one of its kind ever.

For another, the machining requirements are such that several extremely large components—including a giant bridge-type head changer and an even larger bridge-type milling station for the turret opening—had to be specially constructed. Two twin-spindle, 25-HP traveling column machining centers of the 25HC type, and a pair of large, opposed head changers also must be employed to perform all of the required machining steps.

Unlike K & T, which feels it has sufficient systems production space for some time to come, Milacron is crowded. As a result, the company is formulating a series of expansion programs.

Mr. Curtain noted that the recently announced plan to add 57,000 square feet of space to the company's plastics machinery plant in southwestern Ohio will be used to alleviate immediate systems space problems. He declined to give specifics, but did say that Milacron "would be crazy if it didn't build a new plant at some point" that would be totally dedicated to systems manufacture.

Competing head-to-head with K & T and Milacron these days in the systems market is White-Sundstrand headquartered in Belvidere, Ill.

The Sundstrand machine tool operation has been a leader in designing and building ABMSs. It currently has three systems operating in the field, all random, including one at Sundstrand Corp.'s aviation operations, one at Ingersoll-Rand Co.'s Air Hoist Division, and one at Caterpillar for the machining of grader transmissions. It is building a fourth random system for International Harvester Co. to machine components for tractor power trains.

The sale of the Sundstrand machine tool to White Consolidated Industries, and the subsequent loss of some key personnel to other companies, most notably Ingersoll Milling Machine, has set White-Sundstrand back a little, but Thomas Shifo, general sales manager, noted recently that the company is still very alive and very well and very interested in selling the ABMS market.

"Systems are going to be a big part of metalworking in the future," Mr. Shifo said recently. "There is a lot of interest in all segments of the market these days. Will sales reach $200 million in five years? In our view, that's a very reasonable figure."

White-Sundstrand is taking a slightly different approach to the systems market. White-Sundstrand's systems rely almost totally on the company's Modular Series 80 Omnimill machining centers to perform machining operations. These machines are extremely flexible and can perform a wide range of precision machining chores, in part because they can store tool heads up to 11 inches in diameter in their tool magazines and can engage up to six additional tool heads measuring up to 20 inches square.

Because of this capability, White-Sundstrand has not developed an independent head changer.

"We're primarily interested in selling systems that involve a high concentration of machining centers," Mr. Shifo said. "But that doesn't mean we'll shy away from systems that require head changers. If a head changer is required, we'll go to Ingersoll Milling Machine or someone like them for the head changer and we'll build their unit into our system."

The system White-Sundstrand cur-

Variable Mission Automatic Toolhead Changer
General Specifications

Work Surface (Pallet)
Length . . . . 36″ (48″ optional)
Width . . . . 30″ (48″ optional)
Workpiece or fixture . . . . optional

Ranges
Transverse Travel (Z-Axis) . . . . 56″
Rotary Work Index . . . . 360°

Feedrates
Transverse Feed (Z Axis) . . . . 400 imp
Transverse Rapid (Z Axis) . . . . 400 imp
Rotary (B-Axis) . . . . 12 rpm

Toolhead Drive Unit
Horsepower . . . . 50 (input)
Speed Range . . . . 650 to 1950 rpm
Number of Speeds . . . . Infinite

Toolhead Data
Head Size—Width . . . . 36″
Height . . . . 36″
Length . . . . 26¾″

Maximum Tool Length and Holder . . . . 30″
Average Toolhead Weight . . . . . . 4000 lbs.
Hole Pattern, Nominal . . . . 30″ x 30″
Head Storage Capacity Options
  Ready—Access Store . . . . . 18 toolheads
  Bulk Store Capacity (optional) . . . .
  . . . . 18 toolheads
Head change cycle . . . . 20 seconds
Average Cycle Metal-to-Metal . 30 seconds

Maximum Weight Work and Fixture . . . .
. . . . 4,000

Work Index Resolution
Standard 5° . . . . 72 position
Extra Cost option .001° . . .360,000 positions

Pallet Shuttle . . . . . . Optional at extra cost
Chip Conveyor . . . . Optional at extra cost
Machine Weight-less Toolheads
  (approx.) . . . . 90,000 lbs.

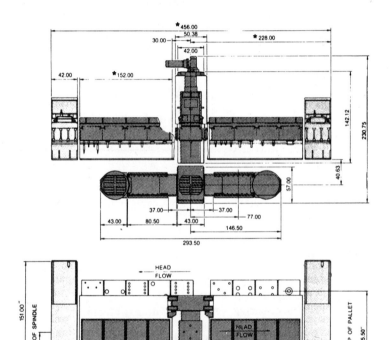

This is Cincinnati Milacron's VARIABLE MISSION automatic toolhead changer. Head changers can also be used as stand alone machine tools.

rently is building for International Harvester gives a good picture of the state-of-the-art at the company. Eight Series 80 Omnimills provide the machining power while a pair of pallet shuttles keep the parts moving.

Unique to the system is the use of a ninth, specially prepared Series 80 Omnimill as the system's inspection station.

"We've taken a little more time in the assembly of this unit," Mr. Shifo conceded recently, "and we've tried to tighten it down wherever possible. But basically it is a standard machin-

ing center with a special probe on it. It works out beautifully because it fits into the system so well."

An unknown factor in the systems market these days is Warren, Mich.-based Bendix Machine Tool, the consolidation of Bendix's old Michigan Special and Buhr machine tool operations.

Bendix has little in the way of actual systems hardware these days, but it does have John J. Biafore as its president, and that may be more than enough to make Bendix a factor to be reckoned with in the future.

Because Bendix has traditionally concentrated on the auto industry, Mr. Biafore is not generally known outside Detroit. But he will be. He has already boosted Bendix's machine tool sales and earnings substantially, and is currently moving forward with plans to double sales in the next three years, from around $55 to $60 million a year now to as much as $150 million a year in 1981.

Bendix's current stockpile of ABMS components consists of one head changing system built primarily from standard industrial components. The company has sold one unit to date, to Cummins Engine Co., for the machining of several different types of flywheel housings at its Jamestown, N.Y. plant. Total cost of the system was well under $2 million.

Yet despite this less than auspicious beginning, Mr. Biafore has an excellent handle on the biggest emerging market for ABMSs of them all: the auto market. In addition, he knows as well as anybody what it takes to sell that market, low prices and high reliability.

"We see a very good market for systems for the machining of parts for the aftermarket," Mr. Biafore said recently. "A lot of these parts are being handled on dedicated equipment and the auto makers are losing their shirts on them. Floor space is critical to these people, and having to carry large dedicated machines for a long time is bad business."

"There's also a very big movement amongst auto makers to get out of dedicated machinery wherever possible," added Donald P. Lamb, Bendix's vice president for special products. "The General Motors Tech Center, here, has spent hundreds of thousands of dollars in recent years developing new ways to break down big dedicated systems into series of smaller, more flexible systems. We think we can play a big role in supplying equipment in this area."

Mr. Lamb notes that it is no accident that Bendix's head changer is designed to accommodate heads used widely on transfer lines.

Mr. Biafore does not apolgize for Bendix's supermarket approach to construction of its head changer. "The systems business isn't going to be all XM-1 tank hull lines," he noted. "We're going after the little guys as well as the big guys, and the little guys need components that consist of parts they're familiar with and that don't involve a lot of exotic solutions to simple problems."

"Price is important," adds Mr. Lamb. "For $400,000, the base price of our unit untooled, we can put some-

body in the systems business."

Mr. Biafore concedes that Bendix will need more than a head changer to compete in the ABMS market. He also agrees that machining centers are more or less essential ingredients to a sound ABMS system, but sees no need for Bendix to reinvent the wheel.

"The machine we're offering now is very expandable," he noted. "We can do a lot more by adding some stations at the front of it, including machining center stations. When the time comes, we will not be at all reluctant to enter into a licensing agreement with a ma-

---

"The fact of the matter is that the technology is there. Now, it's up to the progressive manufacturers."

Charles F. Carter Jr., manager of advanced machine tools systems, Cincinnati Milacron

---

chining center manufacturer. We've already discussed the possibility with several builders." Mr. Biafore would not name names, but an arrangement with White-Sundstrand—which does not manufacture a head changer—is not beyond the realm of possibility.

Perhaps most impressive of all where Bendix Machine Tool is concerned was the company's decision earlier this year to build a head changer for use on its own production floor. The machine is well along now and should begin making chips before year's end.

"We're very anxious to get one of our machines on the floor to see what it can do," Mr. Biafore said. "We've already tooled up for one of our parts, a mechanical drive housing, and we will be adding others as time goes on."

Bendix is not alone in its effort to take some of the dedication out of the high production machinery traditionally sold to the auto makers. Cross, Ingersoll Milling Machine, Kingsbury Machine Tool, and Snyder all are working on development of head changers, waiting stations, and other components that can improve the efficiency and increase the flexibility of dedicated equipment.

One of the reasons for this is the fact that the auto makers have just begun to realize how inefficient

transfer machines really are.

"For years," one auto official said recently, "the industry believed that transfer lines were the most efficient route to take. But recent computer studies have shown that the average transfer line is only around 50 pct efficient. That's really not good enough. They've got to shoot for 80 to 85 pct, and systems look like the best way to get there."

The official noted that Ford is currently trying to limit all transfer machine buys to units of 10 stations or less in an effort to change the way it traditionally makes things.

Amongst Bendix's competitors, Ingersoll Milling Machine and Cross appear to be making the biggest progress. Ingersoll has sold two head changing systems, one to Caterpillar (who else?), the other to Tenneco, Inc.'s J.I. Case Co. Cross also has sold two head changing systems, one to Caterpillar, the other to an unnamed U.S. manufacturer. The company has encountered problems with the controls on the Caterpillar system and has yet to ship it, but sources in the industry say the unit should go out shortly.

Still another important entrant in the systems sweepstakes with big connections in the auto industry is Giddings & Lewis, Inc.'s Gilman Engineering & Manufacturing Co. in Janesville, Wisc. Although Gilman is primarily known for assembly machinery, it is said to be building two large systems at present employing machine tools built by Giddings & Lewis' machine tool subsidiaries. The destination of the two systems is not known.

Perhaps the most interesting aspect of Gilman's presence in the ABMS market is its incorporation of some assembly operations in its systems. Experts in the machine tool industry agree it is only a matter of time before systems extend far beyond chip making. Assembly, welding, heat treating, functional testing, and grinding are just some of the additional operations builders are looking at closely these days. And as pressure for increased productivity mounts, manufacturers will be looking at even bigger and more complex systems to do more jobs than ever before—"to produce transmissions rather than transmission cases," as Cross President Ralph Cross put it recently.

At present, Yamazaki Machinery, through its Florence, Ky.-based subsidiary, is the only serious foreign entrant in the U.S. systems competition.

But it is serious competition, indeed. The company already has sold one of its highly innovative Yamazaki

Machining Systems to Ladish Forge in Milwaukee for the machining of large flanges, and plans to install a second system in its new, 80,000-square-foot manufacturing facility now under construction in Florence.

The Yamazaki system is one of the cleanest and most compact ABMSs on the market today and employs a number of unique features, including interchangeable tool drums, built-in NC rotary tables, and tool heads capable of unusual machining operations, such as broaching and turning.

As the U.S. market for systems grows, the competition from the big Japanese machine tool builders can only quicken. Machine tool officials generally agree that the Japanese are ahead of the U.S. in some aspects of systems construction. They note that such companies as Hitachi Ltd., Kawasaki Heavy Industries Ltd., Honda Motor Co., and Okuma Machinery Works Ltd. have been building and perfecting manufacturing systems for some time now, and they say it is only a matter of time before some of these companies and their vendors look to the U.S. to increase systems sales.

It is interesting to note that U.S. builders see little threat from Europe.

That's because most officials here believe that European ABMS builders, even the East German builders of the innovative Prisma II system now operated by the Fritz Hecker Kombinat in Karl-Marx-Stadt, have taken an academic rather than a manufacturing approach to systems construction.

"The Prisma II system is an important system." one U.S. ABMS expert said recently, "primarily because the East Germans were not afraid to incorporate difficult operations into it, like superfinish grinding. But is it marketable? I don't think so. The return on it has got to be extremely low."

Although interest in systems continues to rise rapidly, pushing deliveries out to two years or more in some cases, ABMS builders are not without problems.

One of the biggest problems involves the traditional financial parameters of the U.S. machine tool industry, which may work well enough in the area of conventional machinery but which pose problems when it comes to systems.

Take the question of proposals. Traditionally, builders have not charged customers for quotations, even though the cost of preparing a

quote can easily run to $300 or more.

But proposals for systems are quite another kettle of fish. "The cost of developing a big system and presenting it to a potential buyer can run as high as $50,000 in some case," one U.S. official noted recently. "That's why we're trying to be very selective in deciding who we will do business with. If a customer is just fishing for information, we're not interested. We have to be very sure early in the game that he means business."

Several officials noted that their companies are making an effort to charge for proposals. But they said those efforts have met with only limited success thus far.

Another financial problem involves

"Our strategy is to build and ship a system on a modular basis."

Paul R. Haas, vice president and general manager of Kearney and Trecker's special products division

---

the tradition, particularly in the auto industry, of not making progress payments for machinery under construction.

"Return on investment is a very important aspect of systems building," Bendix's Mr. Biafore noted recently. "To get into an acceptable ROI area, progress payments may become a necessity in the future.

"Progress payments are a key issue right now. Systems take a long time to design and build, and systems builders must get relief from the large inventory they must carry for a long time. If industry wants the machine tool industry to thrive and grow, they must take some of the responsibility for these major programs."

K & T's Mr. Haas agreed.

"Machine tool builders are really small companies in the scheme of things," Mr. Haas added. "Financing manufacturing systems is difficult. Progress payments are very important."

ABMS builders also have to improve their overall record where installation is concerned. Reports of 18-month start-up and debugging periods are commonplace, and many

manufacturers are reluctant to commit thousands of square feet of much-needed manufacturing space to expensive machinery that doesn't do anything.

"When systems first hit the market, we in the machine tool industry sold them inappropriately, even though we didn't mean to," said Milacron's Mr. Carter. "We were selling the turnkey concept, and rightly or wrongly, that evolved into the idea that we would go into someone's plant, install a complete system, and get it running perfectly without them putting a finger on it.

"That's not the way to go. We need to get as many people in the plant as possible involved in the system right from the start. That can help make the installation smoother, and can shorten the transition."

Another problem is manufacturing's difficulty in justifying the high initial cost of ABMS. In an industry where the one-year ROI is standard, it's hard to think in terms of two-, three-, and even four-year ROIs'.

Still another problem systems builders face is the probability that competition for market shares from non-machine tool companies is bound to crop up sooner or later.

"Right now," one machine tool company official said, "machine tool builders dominate the systems business because the hardware in a system, the machine tools and related equipment, is by far the costliest part of the package." (The figures vary from system to system, of course, but generally speaking, costs of a system can be broken down thusly: machining units, including local NC or CNC controls, 50 pct; tooling and processing, 25 pct; material handling equipment, 10 pct, and control system, 15 pct.)

"But we see the strong possibility of an Allen Bradley or a Westinghouse getting into the systems business as a prime contractor. Control makers, computer makers, and material handling equipment makers are all candidates."

Another official predicted that the machine tool industry's "most likely" competition in the future will come from the user. He noted that Xerox Corp. has been very successful in installing an ABMS in one of its plants, using equipment supplied almost totally by outside vendors.

A final problem ABMS builders must deal with is the need to convince manufacturers that systems are viable in all areas of production.

Presented at CAM '78

The Impact of Microprocessors on NC and Automatic Manufacturing Systems

By D.W. Pritty

1. ABSTRACT

This paper discusses the role that microprocessors will be able to play in future computer aided manufacturing for microprocessors should form the heart of most of the systems that will be required to implement the concept of the completely automatic workshop.

This paper deals with numerical control systems most specifically but mention is made of the characteristics required by other systems in the automatic workshop.

The concept of building a range of numerical controllers using standard microprocessor hardware and software modules is discussed.

2. BACKGROUND

Against an international background of falling profitability and reduction of resources used in the manufacturing industry the concept of the completely automatic workshop using machinery and data processing devices all controlled by microprocessors seems to offer the possibility of providing a new lease of life for manufacturing.

Microprocessors with computing power less than one order removed from that of a mini-computer have been with us for three or four years now.   The immediate attraction of their performance/cost ratio has led to the introduction of a number of products in the computer aided manufacturing area.   The success of these systems will now depend more on software than hardware and many software people have seen in the microprocessor a return to the early days of computing.   Hardware restrictions are again dominant albeit in a technology which is superior by say three orders of magnitude in the basic parameters of size, cost and reliability, fairly stripped down instruction sets are again the order of the day because of the limited numbers of LSI transistors that can be accommodated on the chip and very little support software has been available until recently.

As in the early days of computing much of the programming of these systems has been carried out with a return to assembly code programming. There is a cronic lack of properly developed documenting systems for software.   For instance there is no body in a software house with the same organisational power to enforce standards as the drawing office has in terms of electronic or mechanical hardware.   This together with the difficulty of understanding both the hardware and software documentation of early microprocessors has meant that at the development stage at least 'the product development team often had to be considerably increased to develop a microprocessor replacement to an existing hardware product.

Fortunately the microprocessor manufacturers are beginning to solve these problems. Documentation is becoming better, that is more easily understood, and the products themselves are improving. Microprocessor support chips are being introduced to reduce the chip count and make the microprocessor interfaces with store and peripherals simpler to implement.

The problems of software development have been tackled too. Development systems are now produced by the manufacturers on which application programs can be developed. High level languages have been introduced to aid software development. The purpose of this paper is to review how the current state of development of microprocessor hardware and software should benefit systems in the computer aided manufacturing fields with particular emphasis on numerical control of machine tools.

3.   THE AUTOMATIC WORKSHOP

The concept of the completely automatic workshop when applied in the traditional field in which n.c. has been used is an ambitious one. For here there is a wide variety of work and the batch quantities involved may vary from medium to very small. However it is perhaps this area that offers the greatest potential gain if an automatic system can be successfully implemented.

3.1 COMPUTER SYSTEMS IN THE AUTOMATIC WORKSHOP

Two basic classes of systems are required. Ones that handle the materials, tools and parts themselves and ones that handle the management information regarding scheduling, planning, and part programming.

MANUFACTURING SYSTEMS

These might cover the following main categories:

a)   Material Handling Machines including Robots
b)   Tool Handling Equipment
c)   Numerically Controlled Cutting, Welding and Brazing Machines.

MANAGEMENT SYSTEMS

These might handle the following functions:

a)   Ordering, Scheduling, Workshop Planning and Loading
b)   Part Program Distribution
c)   Component Progress in the Workshop
d)   Machine Performance Reporting - tool wear, breakdowns etc.
e)   Inspection

3.1.1   MANAGEMENT SYSTEMS

At first sight the hardware cost of these would seem to be
considerably reduced by the introduction of microcomputer systems involving
floppy discs and other newly developed components.   However the chief
concern here is the writing and maintaining the application software.
On the hardware side the costs of these systems will be largely made up
of memory and peripherals (some of them special to the system) and software
costs rather than C.P.U. costs.   It is therefore much more important to
adopt an easily programmed microprocessor which supports a suitable high
level language in this area than to seek the cheapest microprocessor
irrespective of the support available with it.

The most complex of the management systems may in fact be implemented
in a mini-computer using bipolar bit slice microprocessors and a considerable
amount of inter-system communication will be required to reflect the
hierarchical structure of the information to be handled.

3.1.2   MATERIALS HANDLING SYSTEMS AND ROBOTS

These have to transfer materials and tools from point to point and
unlike the management systems have to operate in real time.   However in
general terms the complexity of the computing is no more than that of
point to point numerical control systems.

Microprocessors for these systems do not therefore have to be
particularly powerful and whilst it is very desirable that they should
have good high level language support this is not so necessary as for
the management systems microprocessors where there will be a much larger
body of software required.

3.1.3   NUMERICAL CONTROL SYSTEMS

This was one of the first areas in the manufacturing environment
to be automated electronically and it still provides one of the most
demanding environments for the microprocessor both from the hardware
and software point of view.

4.   DESIGN OF NUMERICAL CONTROL SYSTEMS

Historically numerical control systems for machine tools have been
split into three levels of complexity:

a)   Point to point systems
b)   Simple slope and arc systems
c)   Full contouring systems with cutter compensation capability.

Systems of the first and second classes were implemented in hardware and
the third class of system has recently often been a full mini-computer
C.N.C. system.   Because of the demands on processing power imposed by
the contouring process the first microprocessor based controllers were
of the point to point variety only.   One or two manufacturers took

advantage of the microprocessor technology however to introduce powerful part program editing facilities which could be used together with the controller.

4.1   A MODULAR APPROACH TO N.C. SYSTEMS

One of the main hopes made possible by the advent of a cheap processor such as the microprocessor is that the entirely different controller structures which were previously required to meet the different performance categories given above could now be abandoned and a modular system developed to provide a range of capabilities by adding extra modules to the basic unit.   Naturally such schemes are more easily formulated with a blank sheet of paper rather than an existing product and market history but the rewards of producing a modular system with the resulting increased production of the modules and the ability to serve the awkward upmarket requirements simply are significant.

4.2   CONTINUOUS PATH MICROPROCESSOR SYSTEMS

Proceeding from existing point to point systems to continuous path systems puts considerable demands on the computing power currently available in a microprocessor when cutting circular arcs.   Here the problem is to generate within a time period of around 25 mS a new chord step which approximates to the circular arc required.   Several methods of performing this process, known as interpolation, exist.   Figure 1 shows two well-known methods.

4.2.1   FUJITSU SYSTEM

In the Fujitsu system one bit is subtracted (or added) from the i register and then added into the error register.   One bit is then subtracted (or added) from the j register and j subtracted from the error register.   This subtraction is repeated until the error register changes sign.   Each operation on an i or j register corresponds to a single bit of movement in either the x or y axis as appropriate.   In order to achieve high cutting speeds these operations have to be carried out fairly frequently.   A cutting speed of 6 M/min. and a bit size of 2×10^{-6} metres requires execution of the sequence described above every 20 μS. which is very fast for a multi-byte arithmetic operation in an 8 bit MOS microprocessor.

4.2.2   D.D.A. METHOD

Another commonly used technique is based on the D.D.A.   Here each step generates a short chord of a circle.   However the angle step ($\Delta\theta$) must be sufficiently small to allow cos $\Delta\theta \approx 1$ and sin $\Delta\theta \approx \Delta\theta$ to sufficient accuracy to prevent errors building up while a circular arc is being described.   Again this requires very rapid iteration of the interpolation calculations particularly when going round small circles at high angular acceleration rates.   Incorporating continuous path facilities in an MOS microprocessor based system requires new algorithms or the implementation of existing algorithms in custom built LSI hardware.   Fortunately a number of systems exist for performing interpolation in a microprocessor. One system was referred to at a previous PROLOMAT conference. Most algorithms require high speed multi-byte arithmetic capability to perform the necessary multiply and divide operations. With the continuing development of LSI circuits such devices are becoming more available although the extremely long words required in n.c. still pose a problem.

4.3 CUTTER RADIUS COMPENSATION

Market trends however have dictated greater flexibility at the machine tool and the need to cater for varying sized cutters is now an essential requirement. This requires the calculation of new start and finish points for each slope and arc to be cut. These new coordinates have to be calculated in around 100 mS to cater for the case of short blocks. A completely software implementation of these cutter compensation equations to the necessary accuracy would take several seconds on currently available 8 bit MOS microprocessors. One therefore has the choice of including some additional hardware to perform these calculations or of using a bipolar bit slice microprocessor. Another area where modern trends are making further demands on processing power is that of displays. The market is becoming more sophisticated and demanding the flexibility and quantity of data available in a small C.R.T. display rather than the more restricted displays offered in simple systems up to now. Again therefore one has the choice of obtaining the necessary processing power by creating a small multi-microprocessor distributed computing system using MOS microprocessors or of using one bit slice bipolar microprocessor. The exact trade-off economics between the two approaches are quite difficult to define without doing a detailed design of each system. However the multi-MOS microprocessor system seems superior since a single processor will be adequate for simple tasks, and this can be added to in a modular fashion to provide more complicated systems.

A number of established techniques exist for coupling several microprocessors together and the increasing range of MSI TTL bidirectional bus drives and other components does reduce the package count required.

Where only two or three microprocessors are required for the system then one can use the basic difference in memory and processor cycle speeds to allow several processors to interleave on one common program and data memory cycle. This technique is more attractive where memory can not be easily subdivided into separate sections as was the case with core memory. A previous paper by the author described a system based on these techniques.[1] The continued price reductions of field erasable PROM (programmable read-only memory) and semiconductor RAM tends to favour a system where each processor has its own block of program memory (PROM) together with some of its own and some shared RAM for data memory. When each program is working on its own program there is no reduction in speed but some protocol is required to pass data to and from each processor to the common RAM.

The complexity of the data bus organisation can be varied to allow the common data to be passed into and out of the common RAM with the minimum interruption to processing in P1 and P2.   Figure 2 shows a simple arrangement where the common RAM is also the data RAM for P2.   P2 is stopped while P1 transfers data in or out of the common RAM.   Similarly P2 is stopped while P1 transfers data to the RAM.

Figure 3 shows a more complex arrangement where a separate data RAM is provided for P2 and the common RAM only holds data common to P1 and P2. This of course offers an improvement in overall speed at the expense of increased package content.

As discussed above we require some external arithmetic capability to perform the multiplications, divisions and square roots involved in the cutter compensation calculations in the required time.   This external arithmetic capability can be provided in various forms using around 15 packs of MSI logic with a further 10 SSI packs.

5.   SOFTWARE CONSIDERATIONS

We have already discussed how well-proven documentation standards exist for hardware whereas software development in some cases would seem to have almost been carried out in a manner that one can only describe as furtive. The only piece of documentation available from most young programmers seem to be a program listing with a few comments if one is lucky.   Of course often the existing methods of higher level documentation such as flow charts are inadequate as descriptive tools.   Some other form of notation is required to convey information on those processes that occur from translation of the English language specification to the coding sheet is very much required. Fortunately the listing of a well-structured program written in a high level language with good comments is much more readily understandable to a proficient programmer than an assembly code listing.

The choice between whether to use a high level language or assembly code is rather like the choice between standard logic packages and custom designed logic.   If the production volumes are high enough then the extra efficiency of assembly code can be worthwhile.   Product volumes in the computer aided manufacturing area are unfortunately not in this category.   There a suitable high level language is undoubtedly the best solution.   Language theoreticians argue hotly about the advantages of one high level language over another but to extend the above hardware analogy the choice of which high level language to use is equivalent to the choice of which standard logic range to use - important but much less important than the decision to go for a high level language in the first place.

One of the most significant advantages of a high level language is that it allows the system programmer to concentrate on understanding the problem to be implemented in simple terms rather than having to delve into the intricacies of the instruction set of the processor being used.

Naturally assembly code routines will also have to be written to implement time critical areas of program. The methods of linking these routines with the body of the program written in the high level language must be efficient and simple to implement.

5.1 A NEW HIGH LEVEL LANGUAGE FOR REAL TIME APPLICATIONS

As a result of research work in this area in the Department of Computer Science at Strathclyde University Mr. Duncan[2] developed a new language - PL/F - to be a high level language suitable for implementing on a microprocessor performing in real time tasks. Mr. Duncan chose a subset of PL/1 statements which could be compiled easily and could be coded efficiently on a microprocessor with a simple instruction set such as the Intel 8080 or M6800. In addition one or two novel operators have been included together with a number of compiler controls which allow one to choose disjoint areas of store to hold program and data.

5.2 MAJOR CHARACTERISTICS OF PL/F

The basic features of PL/F (and any other high level language) are as follows:

1)   Named Variables (mnemonics can be very useful in preserving understanding of the problem)

2)   Arithmetic and Logical Expressions

3)   Conditional Branching

4)   Looping

5)   One Dimensional Arrays

6)   Procedures (or subroutines)

7)   Literal strings

PL/F performs these operations in much the same way as other modern block structured Algol-like languages. It does, in addition, contain several features worthy of special note.

5.3 THE ADDRESS OPERATORS

The address operator @ allows the address of a variable name to be found.

For instance the statement X := @Y gives as the value of X the address of Y. This can make programming to solve arithmetic equations particularly easy to understand.

Suppose we require to find a value for x in the equation

$$x = i \times a - j \times b + i$$

where x, i, a, j and b are represented by multi-byte variables X, I, A and B. We would write assembly code procedures (or functions) to perform these multi-length arithmetic operations between two parameters and deliver a result. Figure 4 defines the procedures and shows the resultant PL/F code.

5.4 I/O OPERATORS

Another form of address operator performs the inverse function in that it causes the value of an expression to be treated as an address. This operator is particularly valuable when performing memory mapped input output for it means that inputs and outputs can be handled directly in PL/F without the need for assembly code driver routines. The effect operation is best explained by means of an example.

Consider the PL/F statement X := %0409 H;. This statement transfers the contents of memory location 409 Hex. into the byte variable X.

Conversely the PL/F statement %403 H := X; outputs the value of X to location 403 Hex. in memory.

5.5 EXTERNAL MACHINE CODE ROUTINES AND IN-LINE CODE

Simple PL/F statements exist for linking routines written in assembly code in with a PL/F program. Short lengths of machine code can also be inserted into the PL/F program conveniently.

6. CONCLUSIONS

The paper describes techniques developed at the University of Strathclyde to allow the use of MOS microprocessors in modern high performance numerical control systems. The techniques described should allow the development of a range of controllers with increasing degrees of complexity from simple point to point to continuous path controller with a specification similar to previous mini-computer C.N.C. system built up using modular hardware and software.

Once microprocessor technology becomes fully exploited and suitable high level software becomes widely available then the opportunity to produce more intelligent controllers that are simpler to program at little extra cost becomes a practical possibility. This would appear to be a more cost/effective approach than paring the cost of existing controllers to a minimum.

Other systems in the computer aided manufacturing area have not been described in detail but these are likely to present less technically challenging design problems although the problems of integrating these systems into a manufacturing organisation are likely to prove to be very much more significant.

7. ACKNOWLEDGMENTS

The author wishes to thank his colleagues at the University for their help and encouragement in this work. Mention should be made first of the Department of Production Engineering who provided funding to enable hardware to be constructed to prove the interpolation algorithms. Mr. Scott, in particular, provided much helpful advice and gave of his time freely. In the author's own department Mr. Duncan and Mr. Smeed have contributed much to the work that has been carried out.

8. REFERENCES

1. Mr. D.W. Pritty, A Continuous Path Microprocessor Numerical Control System, Advances in Computer Aided Manufacture (editor D. McPherson)

2. Mr. F. Duncan, Work to be published as part of Ph.D. Thesis, Department of Computer Science, University of Strathclyde, Glasgow.

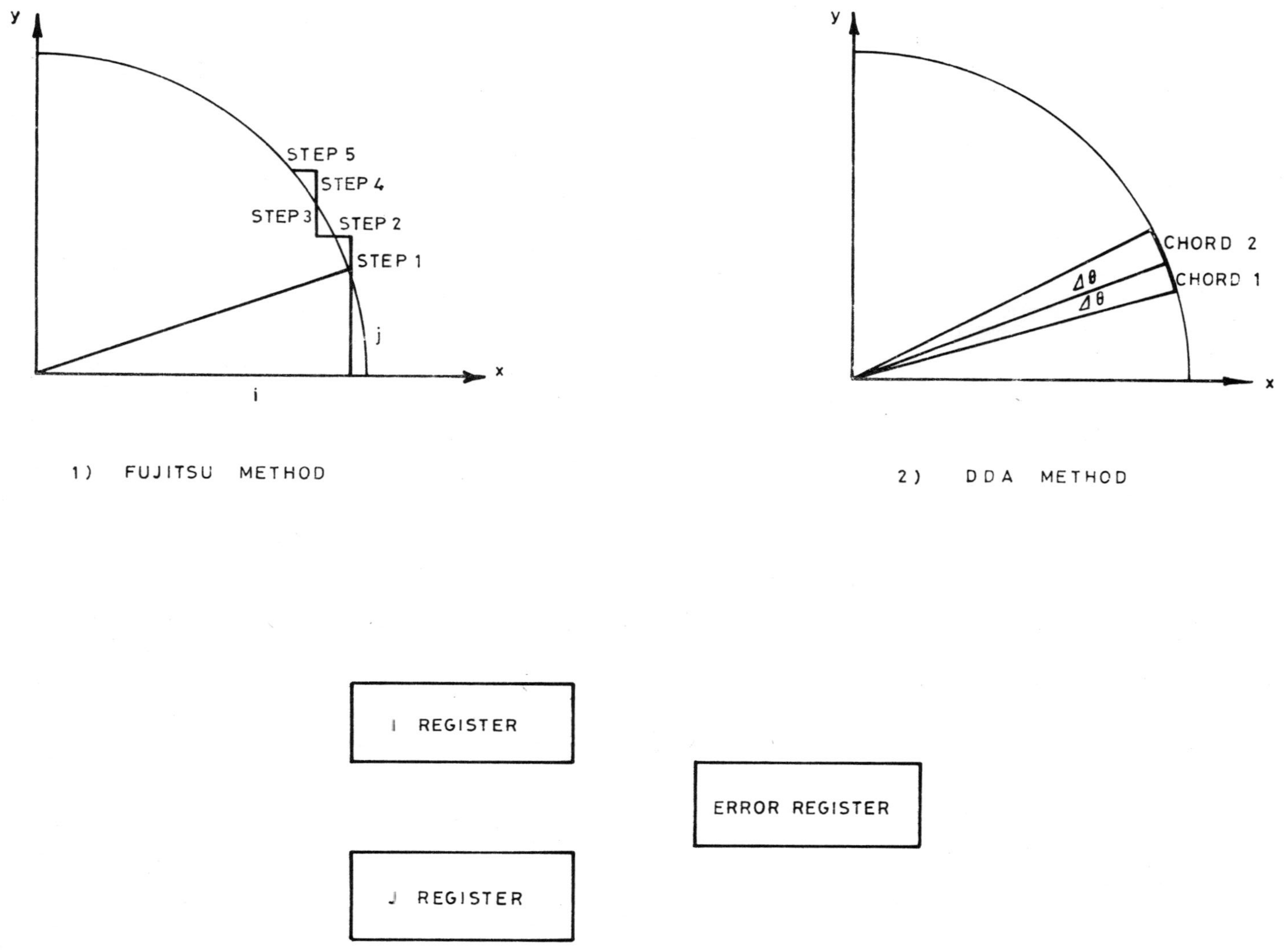

1) FUJITSU METHOD

2) DDA METHOD

FIG 1 INTERPOLATION METHODS (Illustrated for circle centre 0,o)

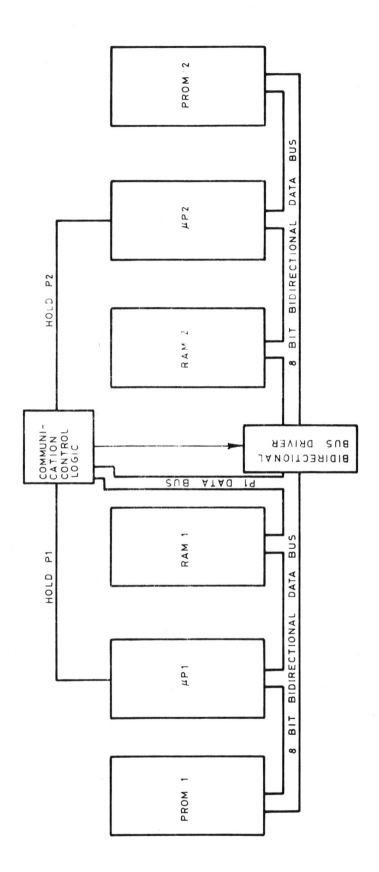

FIG 2   MULTI – PROCESSOR   SYSTEM   WITH   COMMON   RAM

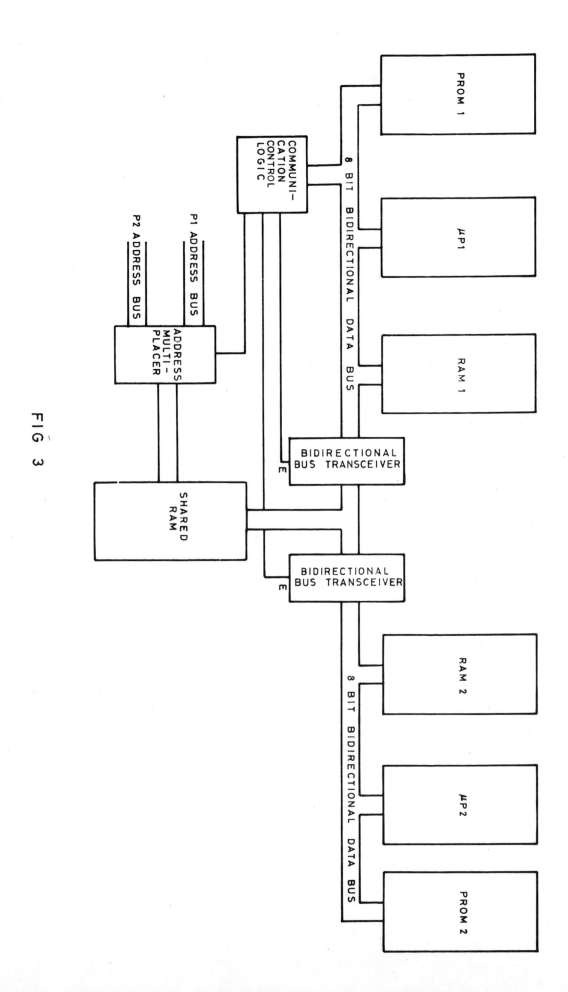

FIG 3

182

CODING EQUATIONS - PL/F - AN EXAMPLE

a) All procedures use two parameters and deliver the result of the operation in the second parameter. The procedures are declared as follows and assembly code routines written to perform them.

```
PROC TRANSFER(@C,@D)        transfers the contents of location C to D
PROC MULT1(@M,@P)           multiplies M by P and delivers the result in D
PROC ADD4(@C,@D)            adds C + D and delivers the result in D
PROC SUB4(@C,@D)            C is subtracted from D and the result delivered
                            in D.
```

b) The following is the assembly coding of a typical procedure SUB4

```
            122   ~
            123   ~
            124   ~      QUADRUPLE LENGTH SUBTRACTION
            125   ~      ------------------------------
            126   ~
2C5C 69     127 SUB4:   MOV     L,C          ~ (DE)[0:3] - (BC)[0:3] -> (D
2C5D 60     128         MOV     H,B
2C5E 1A     129         LDAX    D
2C5F 96     130   .     SUR     M
2C60 12     131         STAX    D
2C61 13     132         INX     D
2C62 23     133         INX     H
2C63 1A     134         LDAX    D
2C64 9E     135         SBB     M
2C65 C3 6C 2C 136       JMP     SI1
            137   ~
            138   ~      TRIPLE LENGTH SUBTRACTION
            139   ~      ------------------------------
            140   ~
2C68 69     141 SUB3:   MOV     L,C          ~ (DE)[0:2]-(BC)[0:2]->(DE)[0
2C69 60     142         MOV     H,B
2C6A 1A     143         LDAX    D
```

c)

```
911 1
912 2       INTERPOLATE: PROC;
913 2           IF MOTION =0 THE
914 2           IF LINORCIRC THEN RETURN;
915 2           IF CHECK() THEN DO; CALL HOME; GOTO CONT; END;
                CALL TRANSFER(@A,@T1);  CALL MULTI(@I,@T1);
                CALL TRANSFER(@B,@T2);  CALL MULTI(@J,@T2);
                CALL SUB4(@T2,@T1);  CALL ADD4(@T1,@I)
                CALL TRANSFER(@I,@X);
                CALL ROUND(@XNEXT); CALL ROUND(@YNEXT);
931 2           CALL RARPART(@XNEXT+3,2); CALL RARPART(@YNEXT+3,2);
932 2           RETURN;      ↓
933 2       END INTERPOLATE
```

Figure 4

Reprinted from Iron Age Metalworking International, January 1979

Kearney & Trecker unmanned machining centres for ASEA's Ludvika plant.

Unmanned Manufacturing A Reality
Unbemannte Fertigung ist Wirklichkeit geworden
La fabrication sans opérateur, une réalité
La fabricación automática es una realidad

By Peter J. Mullins

The factory of the future could well be entirely automatic. Science fiction writers have been telling us for years exactly what it will look like. Rows and rows of near-silent machines—entirely under their own built-in computer control—clicking and whirring away, making pots and pumps, radiators and refrigerators, transformers and TV sets.

In fact, the picture is not of the far distant future. It's just round the corner.

Verschiedene Arten von unbemannter partienweiser Fertigung sind derzeit in mehreren Ländern in Erprobung. Das Endziel ist die automatische Fabrik. Bei einer grösseren Anzahl von Systemen kommen Roboter und andere Handhabungsvorrichtungen für das Laden und Entladen von Werkzeugmaschinen zum Einsatz. Die neuesten Maschinen sind mit Abtast- und Messvorrichtungen mit hochentwickelten Rechnerprogrammen für unbemannten Betrieb ausgestattet. Die Maschinenkonstrukteure tragen dem durch neue Gestaltungen Rechnung. Das Hauptaugenmerk ist auf Bearbeitungsvorgänge gerichtet, jedoch werden unbemannte Techniken auch für viele andere Operationen erwogen.

Plusieurs espèces différentes de systèmes de fabrication sans opérateur, pour fabrications par lots, sont en cours d'essai dans de nombreux pays. Le but ultime est l'usine automatique. Un grand nombre de schémas impliquent l'utilisation de robots et autres manipulateurs pour desserte des machines-outils. Les machines les plus avancées comportent des dispositifs de détection et mesure avec des programmes sophistiqués, pour calculateurs, en vue du fonctionnement sans opérateur. Les concepteurs des machines prennent cela en compte, avec de nouvelles configurations. Alors que les opérations d'usinage font l'objet d'une attention extrême, les techniques de fabrication sans opérateur sont envisagées pour toutes les opérations.

En muchos países se están probando diversos tipos de sistemas de fabricación automática para la producción en lotes. La meta a la que se aspira es la fábrica automática. Muchos de estos sistemas comprenden el empleo de autómatas y otros manipuladores para cargar y descargar las máquinas herramienta. Las máquinas más de avanzada tienen dispositivos detectores y calibradores con programas de computadora muy complicados para el funcionamiento automático. Los diseñadores de maquinas están ideando nuevas configuraciones para ello. Aunque se presta más atención a las operaciones de maquinado, también se están considerando técnicas de fabricación automática para toda clase de operaciones.

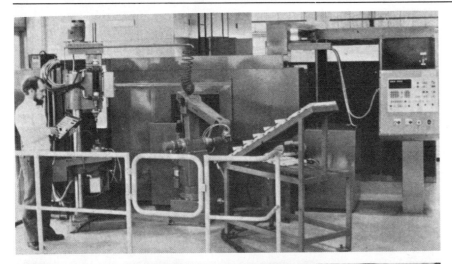

Unusual arrangement for turning and drilling tensile test-pieces at Torshallamaskiner. The robot is behind the lathe, feeding parts through the back of the machine. In the upper picture, the operator is holding the robot's teach-control unit.

Ungewöhnliche Anordung zum Drehen und Bohren von Probestäben bei Torshallamasckiner. Der hinter der Drehmaschine aufgestellte Roboter führt die Stücke von rückwärts in die Maschine ein. Im Bild oben hält der Bedienungsmann das Instruktions-Steuergerät für den Roboter in der Hand.

Un arrangement inhabituel pour tourner et forer des pièces d'essai travaillant en traction, chez Torshallamaskiner. Le robot est derrière le tour, amenant les pièces par l'arrière de la machine. Sur la figure du haut, l'opérateur tient l'unité de programmation-commande du robot.

Una disposición extraordinaria para tornear y taladrar piezas de prueba de la tracción en la Torshallamaskiner. El autómata está detrás del torno, alimentando las piezas por la parte posterior del torno. En la ilustración de la parte superior, el operario sostiene la unidad instructora del autómata.

The first automatic factories are already being designed, and several different kinds of unmanned manufacturing setups are being tried out.

In Japan, the prediction is that such fully-automatic unmanned workshops will be in production by the mid-1980's. One national project there is the ICMSLA project (International Complex Manufacturing System with Laser Application). ICMSLA is currently at the planning stage, and it includes lasers and automatic welding as well as machining.

One design of unmanned workshop in Japan, known as "Superbox," utilizes two machining units, two table-type positioning units and two robots for work loading and unloading. The set-up manufactures

drive units for NC machines and is said to cut floor space by half, cycle time by about 15% and production costs by a similar figure.

In Europe, Sweden, with its very high labour costs (currently the highest in the world), is running hard in unmanned manufacturing in a desperate bid to keep its exports competitive. Most advanced perhaps is ASEA, the electrical giant that has pioneered the technique at several of its factories. In fact, its factory at Helsingborg on Sweden's West Coast gives us a pretty good idea of what the typical medium-sized European manufacturing plant of tomorrow may look like.

Helsingborg is the main manufacturing centre for ASEA's mechanical products' division. It is one of eleven manufacturing plants within the ASEA group; it makes hoisting equipment, an extensive range of industrial gearboxes, forklift trucks and very high-pressure presses.

Under Tore Lindgren, general manager, the plant has begun a calculated swing to unmanned machining, and this involves a complete rationalization of its whole manufacturing system.

Explains Mr. Lindgren: "One of the major problems in making industrial gearboxes of the type we specialize in is the wide range of variations needed to satisfy customer requirements. This makes batch sizes very small—typically about 500 to 1000 a week.

"In the past, a fairly conventional transfer system was used since it was considered that the volumes were too high for single NC machines: There are five sizes of housing and numerous different shafts. This means that changes of tooling were needed frequently and the system was highly inflexible. There was a lot of downtime during toolchanging.

"What we have done now is to put down specialized lines for the smaller and larger gearboxes respectively. The larger boxes are made on a job-shop basis with automation introduced where possible, and the smaller line reorganized into cellular manufacture with groups of machines fed by central robots. In each case we have produced a logical flow system through the shop: Housing manufacture first, then shaft and gear manufacture, followed by heat treatment, assembly, test and stock."

The shaft cell is the simpler of the two ASEA robot-based cells. It comprises a Boehringer PN420 lathe, Sajo milling machine, Nyberg and Westerberg grinder, a Sinico cutting-off machine and a CE Johansson automatic measuring machine.

Slugs are first cut to length on the Sinico unit and centred. The robot then loads the lathe, which carries out a number of machining operations. The part is transferred from the lathe to the measuring unit for the first of a series of inspection checks. The measuring unit contains a built-in microprocessor that controls the lathe in a closed-loop system to keep the dimensions within a set tolerance. If the dimensions stray outside given limits, the system is stopped, and warning signals indicate that attention is required.

After checking, the robot transfers the component to the milling machine and the keyway slot is cut. The part is then transferred to the grinding station for deburring and grinding. A final measuring check follows, and this is carried out on the same measuring station. The grinding operation is controlled adaptively by feedback from the measuring station. Once again, a two-stage measuring system is used.

Excluding setting up, 1-1/2 men are used continuously at this cell for overseeing and loading. This compares with five men to produce the same production rate under the old system. Manual handling operations have been reduced from 36 to only four.

The unmanned set-up for making the housing assembly is a lot more complex. It is really two cells in one, each comprising a robot-fed group of machines. Both cells are linked and coordinated electronically.

The complexities of this set-up arise from the fact that the components comprising the complete housing must first be machined separately, then brought together and assembled for the shaft bores to be machined. This ensures that the various parts make up matched sets with perfect shaft alignment.

In each cell, the robot feeds conventional Burkhardt and Weber multi-spindle automatic bar machines. The plate is then placed on a conveyor and fed from one cell to the other. After a washing process,

the robot assembles the two parts and feeds them to the bore-machining station.

The two robots are linked electronically and therefore work together. They can, however, be stopped separately.

What are the overall benefits of these new-style machining lines?

"Undoubtedly, they are much more flexible," notes Mr. Lindgren. "Apart from a total capacity increase of around 50%, we have been able to cut three to five week lead times to a few minutes. We have been able to reduce stocking costs.

"The system is really having a profound effect on our whole operation. In the last few years we have bought twenty-five new NC machines and thrown out a hundred conventional ones."

The results are noticeable in the Helsingborg factory. The plant area has actually contracted in working floor space during the last few years as production has gone up. Not only are there gaps in the factory where machines have been replaced with smaller setups, one complete shop has been entirely razed, and there is now an open space for redevelopment.

Tore Lindgren points out that this represents a big decrease in invest-

ment and assets in building, stocks and materials. And paperwork. "We have saved 50 000 worksheets a year as our operation has simplified," he notes. "This is on dockets for things like transportation, planning and delivery."

The robot installations at Helsingborg cost around $3 million. Another $3 million has been put into unmanned machining at ASEA's Ludvika factory, farther north. The Ludvika plant makes a wide range of heavy electrical equipment. Like Helsingborg, it has a wide product mix with consequent small-batch sizes. Size of components is generally fairly large so, once again, the largest ASEA robots have been used capable of manipulating loads of 60 kg.

There are basically two installations at Ludvika: One built around robots, as at Helsingborg; the other based on the use of machining centres. This is highly interesting and is claimed to be the most advanced of its type in the world.

Like a number of unmanned machining installations in Sweden, the aim is to provide a system designed to operate without manning for a period of from six to eight hours. In other words, a night shift. In Sweden's affluent society, it is becoming

Cellular manufacturing at ASEA's Mechanical Engineering Div. uses a heavy-duty ASEA robot that transports parts from a lathe, milling machine, grinder, cut-off unit and measuring station.

Bei der Zellenfertigung in der Maschinenbauabteilung von ASEA kommt ein Hochleistungs-ASEA-Roboter zum Einsatz, der Werkstücke von der Drehmaschine, Fräsmaschine, Schleifmaschine, Abtrenneinrichtung und Messstation wegtransportiert.

Fabrication cellulaire à la Mechanical Engineering Div. d'ASEA, utilisant un robot heavy-duty ASEA assurant le transport de piéces d'un tour, d'une fraiseuse, d'une rectifieuse, d'une tronçonneuse et d'un poste de mesure.

Para la manufactura celular en la Mechanical Engineering Div. de ASEA se emplea un autómata ASEA para trabajos pesados para transportar las piezas del torno, máquina laminadora, rectificadora, unidad de corte y puesto de medición.

increasingly difficult to find workers willing to do night shifts. But if the enormous costs of highly-automated machinery are to be justified, the machines must be worked as near continuously as possible.

Hence, while during the two day shifts two or three operators may well be available for supervisory duties, the machines must operate virtually untended for the whole night.

At Ludvika, six Kearney & Trecker unmanned machining centres are installed in line. One MM 800 unit is equipped with a six-pallet work-handling system, while five MM 600's have ten-pallet work handling systems. The pallets are mounted on wheeled trolleys and are pulled round an oval-shaped track with a chain.

The idea is that when the day shift is going off duty, it loads up the pallets fully, thus providing a sufficient component supply for the night's work. In fact, one supervisor is on duty during the night.

The system is designed for a work cycle between 45 and 60 minutes and a wide variety of parts can be accommodated. In some cases, the pallets contain only one part; others contain fixtures having as many as four workpieces (each having cycle times much lower than 45-60 secs).

The whole system is very versatile, since many different parts can be machined at will—even on the same machine. A coding device tells the CNC system which particular workpiece is coming for machining. The machining centres are of course equipped with automatic tool-changers. The large unit has a 68-tool magazine, and the smaller ones have magazines with 52 tools each.

One of the other interesting aspects of the Ludvika installation is that it was installed in 1976 and has been working continuously ever since. The system uses K & T's MK II CNC, and it has tool sensing, coolant-fed tooling and adaptive control of spindle rpm, torque consumption and axis feed rates. It includes K & T's diagnostic communications system (DCS).

The Ludvika installation does not use, as might be expected, a master computer controlling all six machining centres. Instead each centre has its own MK II CNC unit. However, the system is designed to be expanded upward and, if needed, it could be integrated easily into a full-blown DNC system.

At SMT-Pullmax AB, Vasteras, Sweden, machines are now being constructed with the next generation of unmanned manufacturing systems in mind. What this company is aiming at is the completely integrated machining system without the need for a separate robot. The machining systems are based on the company's well-known family of high-productivity Swedturn lathes.

Says Claes Nordstrom, project coordinator for SMT: "One approach to improving productivity on lathe work is simply to use a robot for loading a conventional CNC machine. We believe the better system is to integrate all part handling, chip handling and measuring devices with the machine's own control system. This way, optimum producitivity can be achieved, and the customer has only one control system to concern himself with."

Mr. Nordstrom points out the small-batch problem in Sweden as one of the reasons for this approach.

"A great deal of work is going on in other parts of the world—particularly Japan with highly automated systems," he says. "We get the impression however, that the Japanese are concentrating on high-volume manufacturing. Their idea is primarily to reduce the labour content. Apart from anything else, using a DNC machine with a separate robot having its own control system tends to take up a relatively large amount of floor space.

"We aim both to cut labour *and* improve production rate, using highly integrated, compact manufacturing setups having just a single overall control system."

SMT-Pullmax's smallest lathe, the Swedturn 6, is equipped with an automatic materials handling system, comprising a loading arm and blank magazine, as well as a measurement control station.

The part magazine consists of a conveyor along which a pallet is moved forward. The pallet has seven positions for cassettes that act as workpiece carriers. These cassettes are lifted automatically to the loading arm's grip position.

The pallet normally holds enough workpieces for 1-1/2-2 hours of unmanned operation. However, the pallet can be extended if necessary to hold more pallets, so that the lathe can run unsupervised for longer periods. What also has been done is to link together lathes with an intermediate station for turning over the workpiece. Thus, both sides of a part can be machined.

The Swedturn 20 can also be equipped with an automatic loading device for handling parts up to 60 kg. The loading system consists of a portal beam along which a shuttle carriage with a gripping device moves in and out of the machine. A swing gripper with a pusher can be arranged to grasp a workpiece on the material conveyor and swing it up to the upright position for the gripper of the shuttle carriage.

Another Swedish machine tool maker that has gone deeply into unmanned machining is Torshall-lamaskiner AB, Torshalla, near Eskilstuna. A typical setup is for Bofors AB, Sweden's main defence manufacturer. Here Torshalla supplied a small unmanned cell for making shell cases. The cell comprises two Torshalla S 190 CNC lathes and a Johansson measuring station. Transfer from unit to unit is by an ASEA IRB6 robot from a pallet holding 48 blanks. This is sufficient for a six-hour run comprising one shift.

The robot is fitted with a swivelling double hand. Thus, at each stage, a movement is saved when taking the old part from the machine and inserting a new blank. The complete shell case is finish machined in this cell, with the first lathe using eight tools to rough and finish turn the nose profile and machine the grooves for the copper rings.

At the second station, the robot places the workpiece on a hydraulically-operated mandrel, where the outside diameter is finish turned and a threading and knurling operation carried out.

The measuring station checks a variety of dimensions including diameter length and thread form. Like most setups of this type developed in Sweden, the measuring station is comprised of an adaptive control system with a feedback to the turning machines to compensate for tool wear and other machine drift. However, if the tolerances jump outside preset limits (as would happen in the case of a tool breakdown) the cell is brought to a halt with suitable alarms to indicate the stoppage. The measuring station has its own built-in microprocessor to control this function. Overall accuracy of the system is within microns.

Torshalla has engineered this type of cell for maximum flexibility. It can be operated quite conveniently by hand if necessary—or completely automatically.

Points out Mr. O. Hogberg, Torshalla's marketing director: "The way we have designed the setup is that the robots feed the machines from the back. Thus, if we need manual operation, the man can stand on the other side of the robot and gain better access. In other words, the manual operation is from *outside* of the cellular arrangement while the robot works from *inside*. We have aimed the cellular units at operating a complete nightshift without supervision; but with the flexibility we have built in, a man could operate the machines all the morning and then switch to unmanned operation just for the lunch-hour."

Mr. Hogberg describes an even more sophisticated unmanned setup for preparing tensile test-pieces for Granges-Oxelosund, a major Swedish steelmaker specializing in ships' plate. With the high quality standards needed for shipbuilding plate, two tensile test samples are cut from every plate in the X, Y and Z axes. This means that each plate yields a total of six test-pieces. And these must be dealt with very quickly, since the test information is used to decide whether further rolling is required.

Up to now, this work has required five men operating five conventional lathes and working full time. What Torshalla has devised is a cellular setup with one NC S-200 production lathe and an automatic drilling unit, both fed by an ASEA robot. In addition to the robot, there is a loading station with a cassette magazine and an unloading station with an indexing table.

Since there are three different kinds of tests, A, B or C, which can be carried out on each test piece, the permutations become quite complicated. In this system, an operator decides which particular test is called for, and he prepares a punched card. This card accompanies each test-piece through the whole machining

operation and, when plugged into the NC system, it tells the computer which test is required. At the end of the machining cycle, the robot takes the card from the machine and places it with the test pieces in the rack on the indexing table.

Apart from the one programmer, the complete cycle is automatic, with the robot taking the blank from the cassette magazine, placing it in the drill for centring, taking it from the drill to the lathe and then taking it out of the lathe and placing it in the rack with the card.

Mr. Hogberg points out that apart from the huge saving in labour and machinery, the operation is considerably speeded up, and a more consistent quality achieved.

With the type of setup we have been considering, an important factor is flexibility. More and more, manufacturers are trying to devise the type of production line in which several different parts may be made on the same flowline to obtain benefits of scale.

A good example of the latter is a gear transfer line recently installed at Volvo. This can make five different sizes of gear at 100 pieces an hour. The line includes blank turning, spline drifting, gear teeth forming, etc; but the centrepiece is a numerically-controlled compound machine for drilling oil holes. Formerly, five separate units were needed for this operation, plus intermediate transfer devices.

Designed and built by Torshalla Machine Company (S-644 00 Torshalla), the units are equipped with an NC indexing table and 15 pallets mounted on a track. A twin-chuck handling device loads the index table from the pallets. Twin drilling heads are used for the actual machining.

Why use NC when much simpler programming would have done the job?

"To cut downtime during change-overs," Mr. Hogberg points out. "We could have used a less-sophisticated programming method, but NC is flexible and gives very fast changeover times. With this setup, Volvo finds it can keep much smaller stocks of parts—since a given size of gear can be set up very quickly and economically for batch sizes as small as 50 pieces."

At the Bofors heavy forging plant near Karlskoga, a high degree of automation has been introduced. The

company has, in fact, pioneered the use of robots for handling heavy forging into and out of drop-forging presses and counterblow hammers.

Bofor's forging facilities are the largest in Sweden and, on the conventional forging side, are devoted 80% to the automotive industry. (The company also specializes in precision forging largely for the aircraft industry.)

Currently, the company makes crankshafts up to 300 kg for trucks, and heavy con rods, steering knuckles and gear blanks.

The company's aim has been to eliminate heavy manual operations; cut the effect of noise, fumes and heat on the workers (and therefore hopefully reduce a high labour turnover); increase production; and improve quality. To this end, a number of highly-sophisticated lines have been set up, with perhaps the highest degree of automation in the world for this class of work.

The most impressive is the 50 000 kgm (360 000 ft lb) counterblow hammer line, in which crankshafts and other large parts up to 900 kg can be forged. With the old system, heavy forging blanks were taken by hand from a conventional row of furnaces. The worker manipulated a pair of pincers suspended from a sliding conveyor system, which led from the furnace to the rear of the hammer. The part needed to be maneuvered into and out of the press by hand. It was heavy, hot, dirty work.

Today, a rotary hearth furnace is used to heat the billets; this is loaded and unloaded by automatic devices. The billets are discharged onto a roller conveyor and thence to a descaler, where a transfer device loads and unloads them. They go back onto the rollers and thence to the rear of the counterblow hammer, where they are loaded automatically.

For unloading from the hammer, an electro-hydraulic robot capable of handling 300 kg has been designed and built by Bofors engineers. The robot transfers the finished forging to the trimming press.

In one case, a Volvo crankshaft weighing about 124 kg is being produced. Some 13 to 14 men are employed, and a typical batch would be a die run of about 2000 to 2500 pieces. Without automation, 16 or 17 men would be needed.

On a smaller counterblow-hammer line, gear blanks and wear parts are made in batch quantities. Here a similar degree of automation is provided—except that the robot used to take the parts from the hammer is smaller. It is capable of handling about 100 kg, and it was also designed and built by Bofors. One of the big advantages of this robot is that it does away with the need for a security area, and it thus increases production at least 5 to 10%. Once again, a typical batch would be one die run of about 2500 pieces.

Another line equipped with 4000-ton mechanical presses makes smaller crankshafts up to about 40 kg in weight. Again, this is a highly-automated line on which a vibrator hopper feeds the billets to two high-frequency heaters that operate in parallel. Two parallel conveyors take the billets to one high-speed conveyor. This then takes the billets into the 4000-ton clearing press.

The crankshaft is formed in two stages in a transfer-die system, and it is then transported by chain conveyor to a press that both trims and straightens the workpiece. Finally, an electro-pneumatic robot capable of lifting 50 kg lifts the finished crankshaft from the trimming press and places it on a hanging conveyor system. From here, it travels slowly through controlled cooling bays to a station where the shaft is pushed off the conveyor into bins. This 50 kg robot is electro-pneumatic, and the pneumatic components and circuitry have been provided by Atlas Copco, of Stockholm.

Robot built by Bofors transfers hot forgings from forge to conveyor line.

In all, the line uses only two operators: One for supervising the vibrating hopper and one for supervising the presses. This line cuts heavy, dirty work to a minimum, and it reduces drastically the amount of labour needed.

In Britain, where the urgency of switching to high automation is not so great due to comparatively low wage levels, WE Norton (Machine Tools) Ltd, of High Wycombe has decided to take the lead.

At the recent PEP exhibition in London it demonstrated two of its latest CNC machines that are fed and unloaded by a Unimate robot to form a complete fully-automated production unit for making the centre stem of a marine winch.

Comments Brian Johnson, WE Norton's director: "By harnessing the precise power and versatility of the Unimate we can demonstrate that automated cellular technology cuts lead times and reduces work in progress. Britain lags in this field at the moment; the important thing is that the method not only boosts production and efficiency, it can relieve operators of humdrum work that, although tedious and tiring, still needs great precision and concentration." □

Reprinted from Product Engineering, January, 1978, a Morgan-Grampian publication © 1978

As prices go down, performance up designers accept computer graphics

CRT consoles, not long ago suitable for only the largest companies are now making a way into small ones. And this is only the beginning

In the space of a few years, hand-held calculators have moved from a luxury item for the affluent few to a commonplace machine in just about everybody's pocket. Much the same trend is taking place in computer graphics. Once justifiable only in the largest aircraft and automobile companies, this equipment is now priced in a range attractive to small and medium size companies. And those making the equipment predict that the prices will drop sharply over the next few years.

Thus it seems likely that graphic terminals will be in just about every engineer's environment in the not distant future. At the recent CAD-CAM show in Detroit, IBM announced that it will market Lockheed's CADAM software. Says an observer of that event: When IBM goes into a marketplace it is a big one. And that makes all of us a bit more respectable.

There is a wide variety of equipment available now. Many companies including Computervision, Applicon, Auto-Trol, Calma, United Computing, Information Displays Inc., and Gerber have developed turnkey systems, backed up by software. Others, such as Lockheed, are supplying software. Some new software could possibly revolutionize the way in which designs are made and transmitted downstream to manufacturing. Other companies like Tektronix and Digital Equipment supply key components to the turnkey systems along with a few stand-alone units for design or analytical problems.

Finite element analysis model is prepared on the new Tektronix stand-alone. Finished model is analyzed by large computer

For volume drawings. The cost and performance capability varies widely. Take Lockheed for example. It was one of the first companies to raise graphics to a practical art, developing with in-house technicians the CADAM software that has become a general purpose drafting package with numerical control capability.

Less than two years ago Lockheed began licensing CADAM as a commercial venture. It is being used now by large concerns with a heavy drawing load—auto companies, architectural and engineering firms, and perhaps soon by some heavy agricultural and construction equipment companies.

The software is designed for two different configurations: one centralized where the display terminals are close to the host terminal (depending on equipment up to three miles away) and a satellite system where data may be transferred between a central computing facility and engineering and manufacturing facilities thousands of miles apart.

The system is adaptable to very high volume work. At Lockheed there are almost 200,000 drawings in the data base at any time, with about 5000 on line daily. Normally an individual is assigned space in the data base for a personal drawing file, says George Wicker. The engineer can work perhaps on 100 drawings in this assigned space. When they are finished and engineering management approves them, the drawings are copied from the engineer's file and released into a master file.

The equipment to handle this volume of work is large. For a centralized system, the smallest workable CPU is an IBM 370/135 or equivalent machine from other vendors such as Amdahl, Univac and CDC says Wicker. The scopes, all of the refresh type (explained later), are on the scale of an IBM 2250 which must be within 2000 ft of the host computer or an IBM 3250 which can be a mile away. Other workable scopes include Saunders, Vector General and Adage.

The software comes in nine product modules with up to eight usable for any given configuration. A normal system would be priced in the range of $40,000 with $3800 monthly charges. A very basic system without numerical control would run about $29,000 with a $2700 monthly carrying charge.

Refreshed. Among users of graphics, the advantages of refresh or storage tube scopes is hotly debated. Refresh technology is similar to TV so that changes to a drawing show up almost instantaneously. But the images tend to flicker when there is a lot of information on the screen, users say. Storage tubes must repaint an entire drawing with each change and the wait can

be tedious. Technical advances in the offing, however, may reduce repaint time.

The turnkey systems generally include a minicomputer which will drive several graphic terminals and plotters or digitizers. The hardware is backed by software that enables a user to make two or three dimensional drawings, view them from different angles, make changes in the design, and dimension or crosshatch the parts. Many of the systems also have software for certain kinds of analysis—modeling for a finite element stress analysis or making studies of mass properties, for example.

Each vendor boasts of capabilities that set its equipment apart and each user has a design problem that must be matched to equipment. Often customers shopping

for a turnkey system ask the wrong questions, says William Weksel of IDI (Information Display Inc.). Instead of focusing on a problem and whether the equipment will solve it cost-effectively, too many engineers become concerned with memory size and other specs that may not be relevant to the problem.

As for IDI, it offers scopes that are refresh type, yet are competitive in price with storage tube systems. Among software options, Weksel includes a "family of parts" program that enables rapid design of a new part by simply changing dimensions of a master drawing.

Gerber's David Ryan, who is marketing the IDS-3, claims that his company came up from the mechanical area instead of electronics and so is very sensitive to

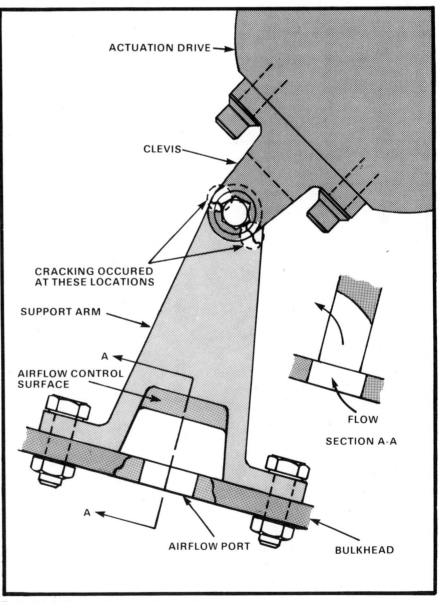

Typical redesign problem solvable on a Computervision system is this support arm and clevis which developed axial-direction cracks in the support arm under high g loading

mechanical design problems. The Gerber system is designed around a Data Management Network in which the graphic system is linked to a large scale computer. The Data Management System helps keep tabs on drawings and, says Ryan, saves a lot of designer time normally spent in saving drawings. Each drawing when completed is given a fixed part number and 250-character description. It is then removed from disk to tape from which it can be easily retrieved months later.

Computervision has designed its own minicomputer to push speed faster. The computer can handle 10 tasks at once. That is it can operate two CRT terminals, a tape drive, NC parts programmer, and up to ten peripherals. Internally they have done away with wire-wrap, and gone solid-state to reduce cooling difficulties.

Small company. The smaller machines based on minicomputers generally do not have the capacity for production drawings but are carving a vital niche in speeding up the design process.

Users adapt graphics to their own needs, product line, and staff. Richard Mattson, senior engineer, Prestolite (Toledo, Ohio), for example, says that its system was installed because the company had a need for much better communication with the designer at the prototype stage to move products into customer hands faster. The company makes large traction motors, most of which are basically the same but differ in some details for each customer. With graphics these modifications are made quickly from data stored on magnetic tape. NC machining tapes are derived from this data base.

In two years of working with his system, Mattson has found that its major benefit has been in allowing mechanical engineers to do creative designs on the screen. They don't have to go through the difficulties of explaining design constraints to draftsmen, a sometimes time-consuming problem. It is not necessary to annotate the layout or include other detail records on such studies. And errors from transposition of numbers and other small slips down the line are fewer.

His software includes a 3-D capability but Mattson uses it only about 1% of the time, when checking tolerances and the interrelation of moving parts.

Large company. At Ford Engineering's Product Development Group, Douglas Miller supervises nine IDS-3 design stations, supported by three host computers, in a system operated by 21 trained designers and draftsmen. Three more are in training. It has been in operation since August.

The system is being used effectively, says Miller, for chassis design work, with some rather dramatic reductions in time. One problem committed to the scope was that of determining flexibility at the linkage that ties the transmission to the car body. The various parts can be simulated on the CRT as the engine accelerates in a way that shows views of the lever action in various positions. This was done in an hour and a half on the computer. Normally it would take a designer 14 to 20 hours at the board.

Infringement studies are another significant application. In an actual case, Miller and his colleagues looked graphically at a new inlet pipe with a flange to see if it would fit and mate properly. The worst set of tolerance conditions were introduced —the maximum dimensions at one place; the minimum at another to see what conditions would prevail.

The real payback, Miller says, is not so much the time saved on drawings but in enabling engineers to look at things that they could not look at before. To make many desirable drawings would simply have been too time consuming but when lines, arcs, circles and other geometrical shapes form at electronic speed and can be changed at electronic speeds, these studies become practical.

Working on catalytic converters on a refresh system a while back Miller looked at a cone in a catalytic converter and found that by angling it slightly, the same cone could fit into several Ford and Thunderbird models. All that was needed were different welding fixtures. This kind of information can reduce part variety and slash costs in large production runs.

In selecting a system, Miller says, it was important for it to have 3-dimensional capability and also what graphic specialists at Ford like to call model space. This means that the x, y and z axes are fixed in space so that the observer always sees the indentation of the part from whatever way he is looking at it. Thus, from a plan view the x and y axes are horizontal and the z shows depressions and shapes; from a side view, the x and z axes are horizontal and the y shows the shapes.

As far as mechanical design goes, Charles Brantly of Hamilton-Standard

Once finite element analysis verified changes to support arm, this redesign of airflow control surfaces was needed to maintain dynamic and static characteristics

runs a graphic center venerable with age. The first scope and a 4-million word disk pack were installed in 1974. Now he has four scopes and a 40-million word disk pack and a plotter. The system is tied into an IBM 370.

One of the areas that allows the greatest cost savings is in design of propeller blades. The theoretical design is entered into the graphic system from the computer as points in three dimensions. When it is shown graphically on the CRT, configuration problems become quite evident. The leading and trailing edges can then be modified if necessary. The trailing edge perhaps has a turning radius that is too sharp to manufacture and must be altered; the leading edge perhaps needs protection against flying objects and birds.

Once the configuration is perfected, drawings are made for manufacturing, quality control and tool design. All these groups use the same data base and have eliminated about 99% of the errors that occur with human transmission of drawing information.

Finite element modeling of blades and wheels is also possible with the system. The data is sent to the large computer for analysis and then displayed on a Calcomp plotter for showing stresses and deformations. If necessary, the model can be displayed again on the CRT to make the grid finer if a closer look is needed at some heavily stressed points.

In ten years, Brantly expects most drafting to be done on scopes with computer data used to make microfilm directly without any need for drawings.

People vs. machines. How well does a technical staff take to such equipment? Most users adapt to scopes easily, though there are people who probably never will. Ford's Miller says that some people simply have a personality clash with the scopes. If one pushes the wrong button, the machine will do the wrong thing. People have to learn to adapt to the machine, stop fighting it and recognize how it works.

Experienced people also find that after a day or two of getting to know the equipment, an operator begins to do productive work. Len Freibott of Crefton says the two design engineers using an IDI system for designing molds and die castings were performing usable work in a week and a half.

Volumes in space. One of the reasons for introducing graphics into engineering and manufacturing is to automate the process, says Richard Judy of Boeing Computer Services. Engineering knowledge should be transferred with a minimum of effort on the part of engineers to detail drawing, test and manufacturing. So far these efforts have failed.

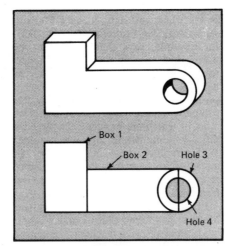

Building from primitive shapes, a complex part can be pictured from computer model

The transmission of engineering knowledge depends on drawings that have to be interpreted and translated back and forth from drawing to computer language along the way.

Judy has hopes the barriers may crumble with a process known as SynthaVision and developed by MAGI (Mathematical Applications Group Inc., Elmsford, N.Y.). "I've been in the computing business a long time and am pretty hard-nosed," he says. "But I am excited about the potential. Only time will tell whether that potential can be realized, whether the software can live up to its inherent ability."

Combinatory vs planar geometry. The SynthaVision software uses a different algorithm to get a three-dimensional view. Instead of putting together an end, top and side view, it assembles primary shapes, such as cones, spheres and cylinders. Once mathematically described as volumes in a computer, the shapes can be displayed on a CRT and photographed with a motion picture or still camera through a color wheel to produce pictures of the shapes that look like the real part.

To build a part, say a simple round ashtray, with these primitive shapes, one would describe three cylinders, the first would be the base of the ashtray, the second the outside rim and the third some-

what smaller would be subtracted from the second to form the open space for holding ashes. These cylinders would be described in x, y and z coordinates.

In all, there are eleven primitive shapes: sphere, right circular cylinder, right elliptical cylinder, truncated circular cone, truncated elliptical cone, ellipsoid, rectangular parallelepiped, right-angle wedge, an arbitrary polyhedron, torus, and a sculptured surface.

Once the part is built by combining geometrical forms in a mathematical computer model, it can be manipulated in many ways and conceivably could serve as the master geometric model for all stages of design and manufacturing. The part can be sliced at different angles to get sectional views, a help in determining if two of the pieces are occupying the same space; it can be exploded with the different parts moved independently; mass property analyses software enable analysts to assign densities to the volumes and calculate weights, the center of gravity, and moments of inertia, and the shapes lend themselves to modeling for finite element analyses.

At Boeing some applications may be forthcoming in the near future, especially for checking numerical control tapes. At present, tapes likely are checked on the cutting machines, often with metal. The waste is obvious if there are errors on the tape. With the computer model, the operation can be simulated with the cutting tools manipulated in relation to the part. If the cutter hits a holding fixture or in some other way performs improperly, this trouble is evident with far less wasted time and materials.

As yet, the Boeing group has not determined whether the MAGI software will produce shapes accurately enough for their exacting demands. And there are the obvious questions about how well engineers who will have to design in terms of geometric shapes will adapt to the different approach. They would have to agree that once the new thinking has been learned and perfected it would enhance the abilities of design engineers.

Other companies are investigating the software, which is available for $150,000 including capability for basic geometry, shaded line drawings, and mass property analysis. Oddly enough, engineering may be the most productive use of a technology that started out as a three-dimensional method for tracking particles through a shield in atomic reactor studies and was commercially adapted to making films for advertisements or industrial training. In a true flight of creative imagination, a Boeing computer specialist saw some films and realized the applications it could have to engineering. And here we are.

CHAPTER 4

TOWARDS CAD/CAM

AD-2000--A SYSTEM BUILT TODAY TO GROW FOR TOMORROW

Patrick J. Hanratty, PhD.
Manufacturing and Consulting Services, Inc.

This paper introduces AD-2000, an interactive graphic system which provides a common base for design, drafting, manufacturing and management information. Included examples show how AD-2000 provides computer independence of programs and data, APT compatibility, both family of parts and macro facility, and complete geometric, drafting and numerical control capabilities. A brief summary of many of the facets of AD-2000 which make it a truly useable system are also included.

INTRODUCTION

From the time of Sutherland's Sketchpad[1] and General Motor's DAC[2], interactive computer graphics has been surrounded by an aura of magic. There has been a sense of expectation and promise that man-machine graphical systems, using the new medium of dynamically changing pictures, might extend man's problem-solving capacities and perhaps even augment significantly the power of the human intellect.

To a large extent many of these early expectations have gone unrealized. This suggests that the approaches taken in building graphical systems have been directed more at "getting out pictures" than at recognizing the true needs of a user in a problem solving environment. Yet, we remain tantalized by the vision that this new dimension of pictorial feedback still holds the promise of aiding in symbolically posing problems and providing a convenient means of getting back solutions.

It has long been recognized that interactive computer graphics is the right medium to reveal properties and solutions in a large body of design areas and that the provision for visual iteration of representations in the design space should ultimately help expedite the generation of good designs.

AD-2000 addresses the problem of finding an effective methodology for rapid synthesis of interactive computer graphic design systems over a wide and open-ended range of application areas. We intend to demonstrate that a significant advance in this regard has been made by the techniques presented here.

What Is A Graphical Design System?

An interactive graphical design system is a dynamic computer medium that helps a user create designs. It does this by dynamic manipulation of design representations under the guidance of the user.

Man builds artifacts to satisfy his purposes. He builds bridges to cross rivers, airplanes to transport people and goods rapidly, and telephones to provide instant communication at a distance. There are many ways to synthesize a given species of artifact satisfying a given set of purposes and requirements. Design is the process of arriving at the specification for the construction of a particular artifact. Because there are so many artifacts that satisfy particular chosen purposes, the search space for the creation of designs is often enormous.

In order for man to contend with these large search spaces he not only requires design representations, he also requires a mechanical medium that will help him manipulate those representations in order to explore a search space rapidly and conveniently.

When design representations involve complex geometrical configurations, as often they do, then computer graphics is the natural medium for their generation and manipulation.

Thus, an interactive graphical design system is a system which allows the user to control the generation and manipulation of graphical design representations, and progressively expose himself to a sequence of such representations that converge on the design that satisfies his needs.

Problem Domains And Representation Domains

Clearly, in a design system, we reason about the construction of real world artifacts through the indirect medium of symbolic representations. These indirect representations, by the nature of computer graphics, are limited to a two-dimensional universe. They are often times pale reflections of the true three-dimensional realities being represented.

Often-times, in choosing a geometrical design representation, we ignore certain features of the real world objects and, for emphasis, exaggerate certain other features.

For instance, in representing a lens, we would not construct the real item, but might instead use two joining arcs to represent a two-dimensional side view, such as given in Figure 1. On the other hand, the path of a ray of light passing through the lens, cannot be seen by the naked eye. Yet we may choose to represent it by an explicitly visible line in the design representation. Thus, we not only have de-emphasized the three-dimensional nature of the lens by using a two-dimensional side view. We have also added to the picture a path that light will follow, which is an intellectual abstraction not directly visible in the real world.

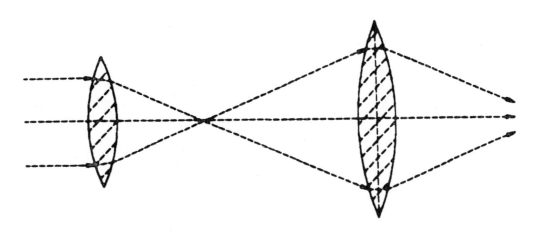

Figure 1

Lens Representation

Thus these representations exhibit two tendencies; they are abstractions that ignore certain information, while adding other information that is helpful in reasoning, even though this added information cannot be seen in the direct visual appearance of the real world artifacts being represented.

However, this is not all. In a design system, we must not only represent objects and abstractions, we must also represent their behaviors and interactions. These behaviors and interactions are captured by mathematical descriptions rendered in the form of symbolic laws. The symbolic laws are, in turn, represented in the computer in algorithmic form. They drive and animate the behavior of the geometric computer representations to allow the consequences of the real world behaviors of real world entities to be seen in the behaviors of the graphical representations. We might picture this view of design systems as shown in Figure 2.

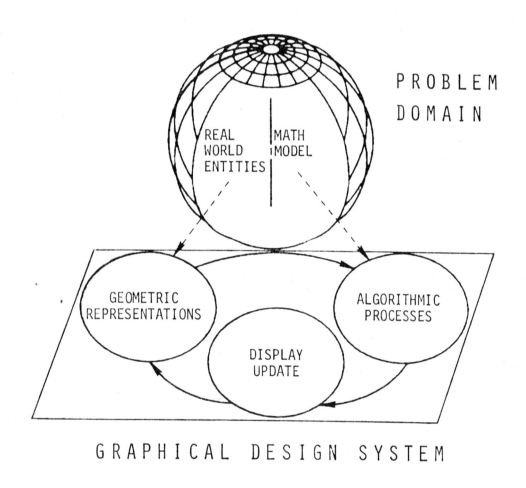

Figure 2

Thus a design representation system must capture both the indirect representation of real world objects geometrically, their mathematical behavioral laws algorithmically, and the interaction between both of these to allow the user to direct the progression of design representations to achieve a design.

AD-2000 BACKGROUND

In September, 1974, MCS concluded work on its mini-computer interactive graphic system, ADAM³, and embarked on a four year design effort. ADAM, which became the basis for a large number of commercial graphic systems, had inherent weaknesses which were not consistent with the future of interactive graphics as MCS saw it.

The four year design effort was a carefully analyzed, structured and scheduled sequence of module developments which would yield what is now known as AD-2000. The design of AD-2000 took into account a large number of factors. Among these was the recognition that:

1. Hardware technology (both computational and display) is exploding, thus machine dependence must be minimized, and even eliminated where possible, so that the system can be moved to cheaper, more efficient computers as they become available.

2. Each major company, and each technological discipline, has its own requirements for geometry, input, output, analytical functions and interface with existing data bases and programs, thus extensibility and "black box" I/0 are manditory.

3. Data (i.e., designs, drawings, etc.) must often be transferred from one type of computer to a computer of a totally different architecture (e.g., the 60 bit word, 6 bits per character CDC 7600 and the 32 bit word, 8 bits per character IBM 370 series) thus data must exist in a form totally independent of computer architecture.

4. Neither MCS, nor any other company, can anticipate the total graphical requirements for any organization, thus growth has to be designed in. An excellent concept so long as the totality of commonly recognized geometric, manipulative and organizational facilities are provided as a base upon which to grow.

5. Graphics must provide a common data base for design, drafting, manufacturing and management information.

6. The existence of significant libraries of APT postprocessors and APT programs, coupled with a large cadre of APT programming personnel, dictate APT input and output compatibility.

Presently AD-2000 is slightly over four months ahead of schedule. The following paragraphs highlight the current status of the system and describes the one area which is under development, with total completion expected by June 30, 1978.

Hardware and Data Portability

Portability, or computer independence can be divided into three distinct, but not necessarily disjoint areas;

1. application software portability
2. data portability
3. I/0 device independence

Application Software Portability

To simplify transporting the AD-2000 application programs from one host computer to another (where different manufacturers, different word size etc. are implied) two steps were taken. First AD-2000 is completely programmed in a subset of ANSI standard Fortran IV. This subset is common to almost all commercially available computers. Second, a computer configuration array (GCA) was established in COMMON. GCA contains virtually all of the parameters which make one computer different from another. These parameters, (bits per word, bits per character, characters per word, number of words per disc sector, etc.), are initialized each time AD-2000 is invoked. Literal values for these parameters are not allowed in AD-2000 except in the initialization routine, thus the application programs are not impacted by a change to a new computer.

The one area which has the potential for causing difficulty in transporting interactive programs from one computer to another is the DATA statement, which provides the definition of the interactive messages. Consistent with the MCS policy, of using a common subset of Fortran for its programs the interactive messages are preset as 2H-- on a 16 bit computer, 4H---- on a 32 bit computer, 10H---------- on a 60 bit computer etc. Since the number of DATA statements which contain messages, menus, questions, etc. exceed 5,000 lines the task of manually converting from one computer to another would be both horrendous, from a time stand point, and could introduce a large number of errors. This particular task is one of the many functions which is totally automated in the MCS text editor (Note: both the MCS text editor and the extensive file management system used in the development of AD-2000 are themselves totally computer independent). A single command converts an entire program from one computer form to another. For instance the command H10, if entered on a 16 bit computer, causes all DATA statements containing 2H to be concatenated into a 10H format. In addition the declaration which established the size of the variable is also automatically changed. Lets illustrate this with a trivial example.

```
16 bit machine statements:
    DIMENSION M1(9),M2(23)
    DATA M1/16,2HIN,2HDI,2HCA,2HTE,2H P,2HOI,2HNT,2HS./
    DATA M2/ 6,2HPO,2HIN,2HT*,2HLI,2HNE,2H*C,2HIR,2HCL,2HE*,2HCO
            2HNI,2HC*,2HSP,2HLI,2HNE,2H*O,2HTH,2HER,2H C,2HUR,
            2HVE,2HS*/
```

```
H10 yields the 60 bit machine statements:
    DIMENSION M1(3),M2(6)
    DATA M1/16,10HINDICATE P,6HOINTS./
    DATA M2/ 6,10HPOINT*LINE,10H*CIRCLE*CO,10HNIC*SPLINE,10H*OTHER CUR,4HVES*/
```

The average time to move AD-2000, to a computer of a make upon which it has never run, is ten working days. This is for a set of programs which consist of over 125,000 executable Fortran statements.

Data Portability

Transporting data from one computer to another is essentially a mechanical process. MCS provides three forms of file save and restore.

1. Computer and COMMON dependent (very fast),
2. Computer dependent and COMMON independent (fairly fast), and
3. Completely computer and COMMON independent (slowest).

MCS has developed the techniques whereby any drawing (designing part, sheet, et al.) may be saved on magnetic tape and then be restored (or merged with an existing design) onto any other computer, regardless of architecture, which runs AD-2000. No source computer identification is required nor do "versions" of AD-2000 impact this total computer independence.

This data portability facet of AD-2000 is obviously important to organizations which make use of many types of computers for the design - drafting - manufacturing cycle, each generating data which must ultimately be integrated into one final design structure.

I/0 Device Independence

I/0 device independence is a matter of programming philosophy. The more modular a system is the easier device independence is to obtain. AD-2000 is structured to allow all current standard I/0 devices and has hooks for new technology. As an example, Tektronix recently announced a 1024 character display buffer for the 4014 storage tube terminal. This refresh facility was incorporated into AD-2000, for message and menu display, in just over two days. MCS has also generated systems in which a combination of storage tube and refresh display were being driven from the same data set under the same operating system.

AD-2000 GEOMETRY

A cursory examination of an AD-2000 menu, listing the basic and extended geometry, make it apparent that the majority of commonly used geometric entities are available, in virtually all standard definition forms, in AD-2000. Figure 3 lists the AD-2000 geometric entity types, without delineating the multiplicity of forms for defining each type. It should be apparent, to those who follow bounded geometry development, that a large number of the extended geometry types in AD-2000 today are curves, surfaces and solids which are being investigated and researched for possible future development, perhaps culminating with success in three to five years, by some of the largest international computer aided manufacturing research groups.

Figure 3 also illustrates a potpourri of AD-2000 curve and surface types to show the flexibility available in surface representation. The non-monotonicity of the AD-2000 cubic spline is apparent in the script "Stratford upon Avon" which required five splines.

While MCS feels that the AD-2000 geometry represents the majority of commonly used geometric entities the system was still designed with the attitude that these forms must be extended for specific users. Thus MCS has generated a "common link" through all curves and all surfaces. This common link is found in the bounded, parametric way curves and surfaces are defined. The consequence of this far exceeds the apparent significance. Programs which determine curve or surface intersections, tool paths, in fact a myriad of functions, are not concerned with what type of curves or surfaces they work with. They are only concerned with the fact that each curve has, as the initial information in it's definition, the beginning and ending parameters which define the extent of the curve. Similiary each surface is expected to have its extent defined by a minimum U, minimum V, maximum U, maximum V as the parameters which are the initial four values found in it's mathematical definition. Thus if company XYZ desires to add its own surfaces to AD-2000, they will be compatible with all AD-2000 functions so long as the evaluator for the surface works on parameters. This concept is not new but it's astounding how few systems take a "common link" approach to geometry.

Application extensibility follows naturally from the context free data form of AD-2000. There are no built in constraints defining what applications will eventually exist or how the applications inter-relate. Since AD-2000 is essentially a three-dimensional system it is conceivable that all applications would require a 3-D data. Fortunately this is not the case. No burdensome overhead is required from unnecessary functions or data. As an example a purely 2-D application does not require any 3-D data or 3-D functions. It was natural that Alameda County, California, USA, selected AD-2000 as a base upon which to build their automated mapping system, a totally 2-D application.

MODALS AND FONTS

All system and application parameters are under the full control of the user as a set of modal values, i.e., values which, once set, remain the same until the user changes them. These modals are divided into three catagories; system, drafting and numerical control. As AD-2000 grows and new applications are integrated into the system the new catagories of modals will be added.

System Modals

System (i.e., not oriented toward a specific application) modals include:

> display or non-display of menus (a prompt with the menu name is always displayed),
> re-enter a specific construction after each definition or go back to the top level of control,
> curve display tolerance,
> curve font, i.e., light, normal or heavy combined with solid, dashed, phantom or centerline,
> surface display control, i.e., number of paths in each direction and number of points for each path,
> number of guaranteed decimal places of accuracy,
> legend display control,
> horizontal and vertical graph scale length,
> tick mark length, and
> data graph title and scale locations.

Drafting Modals

> character size,
> witness line suppression control,
> text and arrow placement control,
> automatically generated or keyed-in dimensions,
> cross hatching material (all ANSI standards except concrete),
> number of decimal places to display with dimensions,
> fraction (from 1/64 to 63/64) or decimal place display,
> label and dimension origin (screen position, delta position, existing point or automatic),
> arrowhead alignment (automatic or manual),
> drafting scale factor,
> character set control (fast, standard, leroy or user generated),
> character slant angle,
> arrowhead length, and
> leader line distance.

Numerical Control Modals

> surface feet (millimeters) per minute,
> milling path display mode (tangent, centerline or both),
> cutter side (+X, +Y, +Z, -X, -Y, -Z),
> coolant,
> spindle direction,
> feed rates (rough and finish),
> spindle speed,
> tool engage mode,
> tool retract mode,
> rough cut distance,
> clearance and retraction planes,
> tolerance (intol and outol for rough and finish), and
> deep hole and chip relief parameters.

DRAFTING FUNCTIONS

The AD-2000 drafting functions provide a sophisticated, automated, drafting environment. They are designed to perform the mundane portions of the drafting activity in a fraction of the time, and with greater accuracy and repeatability, than has heretofore been attainable with manual methods. Figure 4 lists the functions associated with mechanical drafting and presents examples of various AD-2000 drafting capabilities. In addition to these, AD-2000 has a large variety of drafting functions designed specifically for map making, electrical, and architectural drafting.

Labeling and dimensioning provide an excellent example of effective interactive graphics. Man and computer, working together, each performing the tasks for which he and it are best suited, obtain maximum man-machine efficiency. The man can use his judgement to adjust final label positions, special notes, and indicate original dimension positions, all of which would require a large amount of computer time. The computer, meanwhile, performs the tedious chores of computing and centering distances, generating dimension lines, witness lines, and arrowheads, and storing and reproducing data.

NUMERICAL CONTROL

The AD-2000 N/C capabilities are summarized in Figures 5 through 14. What these figures don't show is the degree of control the user has over every phase of machining, including editing (insert, delete and replace) of the final CLFILE. In addition to the individual functions, the AD-2000 N/C modules contain the following:

1. Machining time calculation (approximate) is displayed on demand for every complete tool path, or all paths generated, and printed on CLPRNT for each motion.
2. Any post processor command can be inserted, for inclusion in the CLFILE, after each complete tool path.
3. Circular interpolation is used, if controller has such a provision, for any Pocket, Profile, or Lathe path.
4. The CLFILE can be modified by editing (i.e., inserting, deleting or modifying) the CLPRNT.
5. Any curve (or family of curves) can be projected to any surface (or family of surfaces). The resultant can be machined in either the 3-axis or 5-axis end cutting modes. Tool can be ON, LEFT or RIGHT of the resultant, either normal to the part surface or parallel to the drive surface.
6. The user can control a PAUSE after N cut vectors are generated. The options to the user are then:
    1. insert tool motion,
    2. proceed for N more cut vectors,
    3. proceed until region is complete,
    4. terminate but save path generated, and
    5. terminate without saving.
7. The cutter outline (either milling or lathe cutter) can be displayed (in a refresh buffer on a storage tube) and discrete positioning of the cutter is allowed. At each point where the user desires to define the end of a cut vector he can do so.
8. Any displayed cut vector can be selected, at which time insertion, deletions or modifications can be introduced to the tool path. These functions are controlled by either the CLFILE/CLPRNT edit, by integrating with item 7 above or by use of CANON.

Figure 6 illustrates the AD-2000 Pocket function. The object is to clean out a region bounded by a set of curves. The curves are indicated by the user, who supplies the machining instructions (i.e., number of base and side rough cuts, tool description, direction of rough and finish cuts, coolant and spindle parameters, etc.) after which the cutter paths for cleaning the pocket are automatically generated. Notice that the path closest the actual boundary is very smooth, while the other paths are quite coarse. Only the path where the final side and base are generated has to be machined slowly and with a large number of steps to insure a smooth finish.

Figure 10 is a simplified view of what actually takes place in a design-fabrication sequence. The generation of meaningful examples of AD-2000 machining facilities is difficult, in that tool paths are seldom meaningful except to part programmers, and others involved in metal cutting and tool path generation. The difficulty is compounded when examples of surfaces intersecting other surfaces are required to explain a particular facility such as "region" definition. For this reason, i.e., clarification, curves are being used in the region definition. In practice, the drive/check entities can be other surfaces, surface edges, surface intersection curves or any constant coordinate value. Figure 10 shows the cutter tangent path in four views. Ordinarily, a surface definition extends beyond the actual useable area so cutters move from one surface to the next in producing a part. This is the APT Part Surface, Check Surface, Drive Surface concept.

Functional APT Compatibility

To be functionally compatible with APT could imply that actual communication with APT is desirable. While this may or may not be the case, AD-2000 has been designed considering not only APT functional capabilities but also APT input and APT output.

AD-2000 has, as its standard input, a series of actions by a user at an interactive CRT. The alternate AD-2000 input (referred to as GRAPL) provides for a language which has APT as a subset.

APT output compatibility is provided in four ways; CLFILE, symbolic APT, CLPRNT, and standard APT post processor. The AD-2000 CLFILE is the APT CLFILE. (Actually, the AD-2000 CLFILE is several APT CLFILEs since the major APT systems don't all communicate with each other.) Figure 15 shows how the AD-2000 associative data structure was used to generate symbolic APT statements from graphically generated data.

The CLPRNT output of AD-2000 not only displays the full output but allows the user total edit control, i.e., the ability to insert, delete or replace any output item, or group of items.

MANAGEMENT INFORMATION

AD-2000 supplies a variety of tools to assist management in organizing and evaluating the efficiency of the design/manufacturing processes. These tools include attribute management and data graphs.

Attribute Management

Attribute management has three modes: create, interrogate and delete. The create mode allows the attachment of any form of descriptive information to an AD-2000 entity. This descriptive data exists as an attribute which is always a character string, and sub-attributes which consist of any combination of character strings and/or numeric values. Entities may have any number of attributes.

The interrogate mode allows the user to extract attribute information in any form. Currently the interrogate mode is menu driven but development is in process for a natural language interrogation. The interrogate choices are:

1. Retrieve all entities with specified attribute or sub-attribute.
2. Identify the entity which has a "minimum attribute or sub-attribute".
3. Identify the entity which has a "maximum attribute or sub-attribute".
4. Find the total value associated with all occurances of a specified attribute or sub-attribute.
5. Retrieve all entities with specified attribute or sub-attribute which is "relational" than a specified value. For "relational" read: less than,

$$\text{less than or equal to,}$$
$$\text{equal to,}$$
$$\text{not equal to,}$$
$$\text{greater than or equal to, or}$$
$$\text{greater than.}$$

6. Display all occurances of a specified attribute.

Figure 16 illustrates the sequence for "retrieving a list of all vendors who sell stainless cams with a perimeter greater than 450 mm".

Data Graphs

This function provides for the generation, naming, filing and recall of graphs and histograms. The types of data presentation are:

1. Histograms (horizontal or vertical),
2. Linear plots (with variable X and Y scaling),
3. Equation plots (i.e., evaluation of GRAPL expressions) with up to three independent variables,
4. Polar plots,
5. Logarithmic plots, and
6. Pie plots.

The user can control titles, labels for each axis and curve, text size and angle, tick marks, etc.

The Graph modes include: the capability of defining a graph, a graph template (i.e., a complete, reusable graph sans data), modifying graph data and hardcopy.

DESIGN

The design segment of AD-2000 is the one unfinished area. When completed the GRAPL language, which is the main design tool, will contain all of the facilities required to generate a complete design system for any application. The current status of GRAPL provides for generating either single variable expressions or programs which contain dimensioned variables, GO TO's, conditionals, statement labels and functions. Results of the GRAPL program can be used anywhere AD-2000 asks for input data.

AD-2000 also provides a family of "canned" analytical design aids for both 2-D and 3-D analysis and special analysis for splines.

Spline Analysis

The following analysis is available for any 2D spline:

1. SLOPE
2. CURVATURE
3. RADIUS OF CURVATURE
4. X vs. PARAMETER PLOT
5. Y vs. PARAMETER PLOT
6. EXTENDED ANALYSIS

2-D Section Analysis

The following analysis is available for any closed geometric figure:

1. LENGTH OF PERIMETER
2. AREA
3. CENTER OF GRAVITY
4. FIRST MOMENT
5. MOMENT OF INERTIA
6. RADIUS OF GYRATION
7. POLAR MOMENT OF INERTIA
8. POLAR RADIUS OF GYRATION

3-D Analysis

The following analysis is available for closed figure rotation about a principal axis (i.e., a solid of revolution) or projected along an axis normal to the view plane.

1. SURFACE AREA
2. VOLUME
3. WEIGHT
4. WEIGHT/UNIT LENGTH
5. FIRST MOMENT OF MASS
6. CENTER OF MASS
7. MOMENT OF INERTIA
8. RADIUS OF GYRATION
9. SPHERICAL MOMENT OF INERTIA
10. SPHERICAL RADIUS OF GYRATION

MISCELLANEOUS CAPABILITIES

AD-2000 is too large and complex a system to provide a detailed explanation of each facility in a short exposition. The following sections highlight selected features which contribute to the high level of useability found in AD-2000.

Macro

The dictionary definition of "Macro" is "largeness or longness in extent, duration or size". However, the computer world has adopted macro to mean either "a single computer instruction that stands for a given sequence of instructions" or a "named set of statements which expect certain variables to be preset in order to form a useable definition".

AD-2000 has extended the macro concept one level further. Thus, an AD-2000 macro can be:

1. A MACRO--TERMAC sequence in a GRAPL program.
2. A GROUP, which consists of geometric construction entities gathered together for treatment as a single entity. Groups can be nested up to seven levels deep. Each Group allows an unrestricted number of entities by subgrouping entities.
3. An Array, which is a facility that allows the mass generation of geometric entities in either a Rectangular or Circular Array. Rectangular Arrays include the option of being defined at an angle with respect to the positive X-axis. Both Rectangular and Circular Array allow for exclusion of specified instances of the array if most of the array is to be used, or inclusion (i.e., do not generate instances not specified) if the array is sparce. This exclusion-inclusion capability is defined by column and row number for Rectangular Arrays

and by instance number (counting counter-clockwise) for Circular Arrays. Arrays handle all curves and geometric entity types including Groups and other Arrays. After an Array has been generated, it is possible to operate on an entity created, including the deletion of the base entity.

4.   A PATTERN, which is a set of geometric entities combined to form a complex entity pertinent to a specific application. As patterns are created, they are stored in a library, and are available from a menu selection. PATTERN is a logical extension of the GROUP, DUPLICATE, and MIRROR functions. It provides the capability for defining complex construction elements that may be employed by all users of the system. Thus, standard parts may be defined once and then employed by the designer with the freedom of a template. At the time of retrieval patterns may be placed anywhere on the drawing and may be magnified or reduced in size as desired. Thus, if a single object is to be used several times with only size, location, and rotation differences, it need only be constructed once. A placed pattern can become a set of disjoint entities or can be treated as a group.

5.   A FIGURE, which is the same as a PATTERN except that FIGURES always maintain associativity with their ancestor. Thus, if an original FIGURE is changed, that change occurs in all descendents of the original FIGURE.

6.   A TEMPLATE, which is also similar to PATTERNS and FIGURES except that TEMPLATES have no view association. Thus, a configuration filed as a template will have the same appearance upon retrieval regardless of original or retrieval views. All other features of patterns pertain equally to TEMPLATES.

Levels

Levels can be used to add new entity types to the AD-2000 data structure. These new entity types can be named (i.e., have up to 30 characters as a descriptive identification). Levels can be listed by name and number and treated as entities for blanking, unblanking, translating and duplicating. AD-2000 provides for up to 4096 distinct levels.

Entity Selection Modes

The availability of a rich set of construction features is only significant if the displayed entities can be easily selected for manipulation, interrogation and analysis. AD-2000 provides a variety of means of simplifying the entity selection process. Among these selection modes are:

1.   Individual selection (either light pen, cursor or tablet pick) with momentary lockout. Momentary lockout is the capability of the user selection from only one entity type even though the system allows a number of entities to be selectable. This feature simplifies selection in a cluttered display.

2.   Chaining is a facility which provides that any time a series of entities that are to be selected are contiguous (such as in grouping, patterns, mirroring, cross-hatching, etc.) the user need only select the first entity, indicate a search direction, and the graphic system will automatically locate all of the entities in the contiguous set. This feature reduces the time required for many of the graphic macro operations.

3.   Select by name. Since each AD-2000 entity has a minimum of two distinct names, its creation sequence number and its pointer, selection by name guarantees retrieval of the correct entity immediately.

4.   Select all entities within a circular or rectangular region.

5.   Select all entities outside a circular or rectangular region.

6.   Select all entities on a given level, or set of levels.

Automatic Scaling

No restrictions are placed on the actual dimensions of the final work piece. All information is automatically scaled (if the user so desires) to fit the display screen as geometric entities are defined. This automatic scaling provides total freedom for the development of complex representations without the need for consideration of boundaries or scale factor.

Windowing

Automatic scaling is complemented by the ability to magnify any selected area of the construction represented by the display. Resolution can be obtained to any degree thus allowing entities to be displayed on a scale where details can be easily created, modified and viewed. The display becomes a window providing access to an essentially unbounded and virtually unlimited work area. All data outside of this window is automatically scissored.

View Manipulation

All construction is considered three-dimensional. The data on the screen can be viewed in any orthographic projection or rotated to any auxiliary view that may be desired for a "look" at the construction.

Multiple Views

The user can layout the screen just as he would a drawing, thus up to 32 views can be simultaneously displayed. An entity created in one view is automatically projected into the other views unless the user restricts the views he wishes the entity to appear in.

Translate and Rotate

Entities can be placed in any desired position through the translation and rotation functions of the graphic system. Planar entities can be rotated in the plane of definition, non-planar entities can be rotated about any point. Translation is three-dimensional. Both translation and rotation allow scaling as an automatic procedure in the operation.

Duplicate

Any entity, or group of entities can be duplicated, prior to either a translation or rotation, and then be mass produced by a combination of "Step and Repeat" and "Do N Times" facilities. Each entity of the duplicated objects can still be accessed as an individual entity to allow variation in some of the duplicated items. Duplicated items can also have a scale factor applied to them.

Blanking

The blank mode provides the capability to cause the nondisplay of entities until the time they are required, and thus unblanked. Entities may be individually selected or may be selected by entity type or level number (where all entities of the selected entity type or level are automatically blanked or unblanked). Entities may also be blanked by sequence number, pointer, or all inside a rectangle defined by diagonal points.

Canon

Canon displays and allows modification of the definition (canonical) form of all entities. The canonical form modification consists of selective delete, replace or insert by the user.

Mirror

Mirroring provides reflection of geometric entities about a line as an axis. The mirror mode is continuous and allows the sequential selection of individual entities from the screen or the automatic chaining of contiguous entities.

SUMMARY

With the completion of GRAPL in June of 1978, AD-2000 will provide a total system for design, drafting, management information, and manufacturing. In its current form, i.e., sans the complete GRAPL facilities, it is still a rich and powerful system which provides a user with the most comprehensive drafting and machining system available today. The computer independent aspects of AD-2000 also assure the user that work generated today will not be made obsolete by an ever-changing and expanding computer marketplace.

REFERENCES

1 Sutherland, I. E., "Sketchpad—A Man-Machine Graphical Communication System", AFIPS Conference Proceedings 1963 Spring Joint Computer Conference, V. 23, Spartan Books, Inc., 1963.

2 Jacks, E. L., "A Laboratory for the Study of Graphical Man-Machine Communications", AFIPS Conference Proceedings 1964 Fall Joint Computer Conference, V. 26, Spartan Books, Inc., 1964.

3 Hanratty, P. J., "Computer Aided Design, Drafting and Fabrication on Low Cost Systems", SHARE XLII Proceedings, March, 1974.

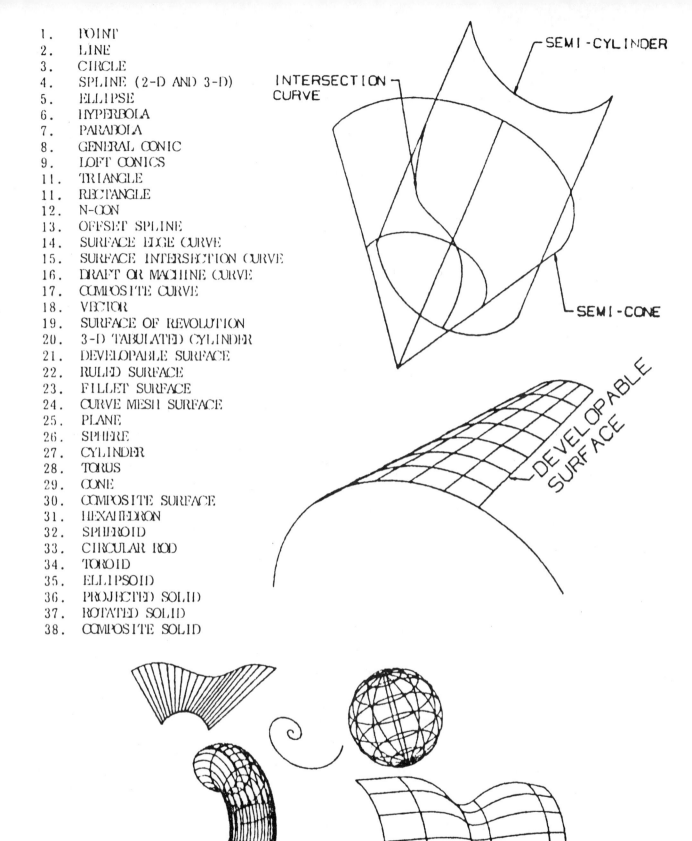

1. POINT
2. LINE
3. CIRCLE
4. SPLINE (2-D AND 3-D)
5. ELLIPSE
6. HYPERBOLA
7. PARABOLA
8. GENERAL CONIC
9. LOFT CONICS
11. TRIANGLE
11. RECTANGLE
12. N-CON
13. OFFSET SPLINE
14. SURFACE EDGE CURVE
15. SURFACE INTERSECTION CURVE
16. DRAFT OR MACHINE CURVE
17. COMPOSITE CURVE
18. VECTOR
19. SURFACE OF REVOLUTION
20. 3-D TABULATED CYLINDER
21. DEVELOPABLE SURFACE
22. RULED SURFACE
23. FILLET SURFACE
24. CURVE MESH SURFACE
25. PLANE
26. SPHERE
27. CYLINDER
28. TORUS
29. CONE
30. COMPOSITE SURFACE
31. HEXAHEDRON
32. SPHEROID
33. CIRCULAR ROD
34. TOROID
35. ELLIPSOID
36. PROJECTED SOLID
37. ROTATED SOLID
38. COMPOSITE SOLID

Figure 3   AD-2000 geometric entities

1. PROJECTED ENTITY
2. CROSS-HATCHING
3. HORIZONTAL DIMENSION
4. VERTICAL DIMENSION
5. PARALLEL DIMENSION
6. ANGULAR DIMENSION
7. CIRCULAR DIMENSION
8. DIAMETER DIMENSION
9. GENERAL NOTE
10. GENERAL LABEL
11. CENTERLINE
12. MODIFY DRAFTING ENTITY
13. DETAIL MAGNIFICATION
14. BALLOON
15. TRUE POSITION SYMBOLS
16. ARROWHEAD AT END OF LINE

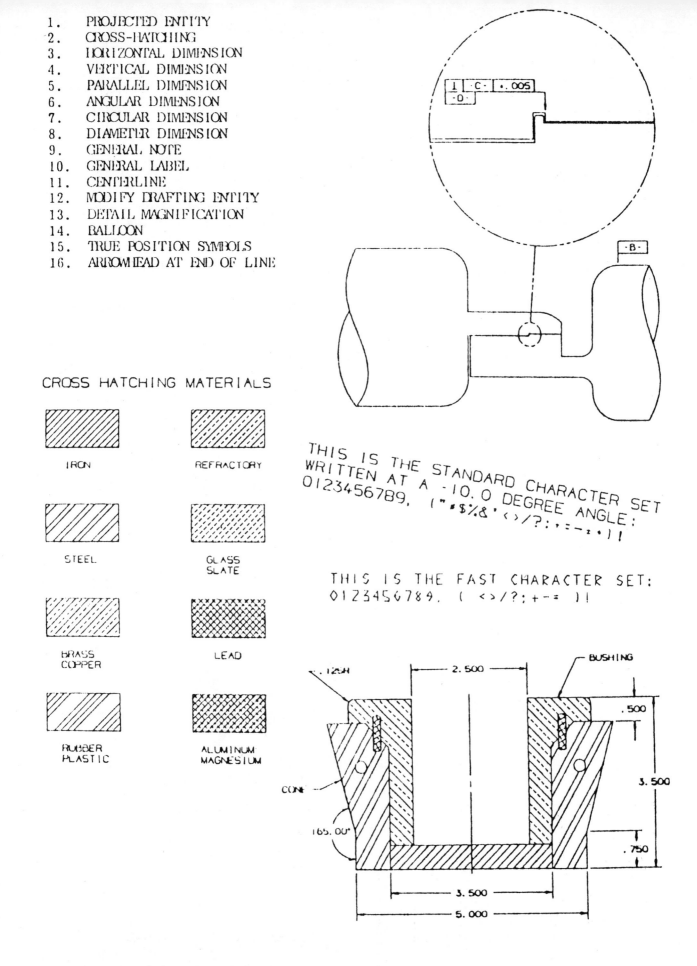

CROSS HATCHING MATERIALS

IRON

REFRACTORY

STEEL

GLASS
SLATE

BRASS
COPPER

LEAD

RUBBER
PLASTIC

ALUMINUM
MAGNESIUM

THIS IS THE STANDARD CHARACTER SET
WRITTEN AT A -10.0 DEGREE ANGLE:
0123456789. ("#$%&'<>/?;+=-=•)!

THIS IS THE FAST CHARACTER SET:
0123456789. (<>/?;+-=)!

BUSHING

CONE

165.00°

2.500
.500
3.500
.750
3.500
5.000

Fig. 4 Drafting functions

212

The AD-2000 Pocket/Profile
functions provide:

1. User choice of simple, moderate or complex analysis.
   Complex handles most cases
   of concavity, notches and
   cusps.
2. Bottom of pocket can be any
   depth or any plane.
3. Capability to insert secondary feedrate commands for
   cornering feedrate control.
4. Full control of number of
   side rough cuts, base rough
   cuts, cut direction and
   entry retraction modes (as
   well as feeds and speeds).

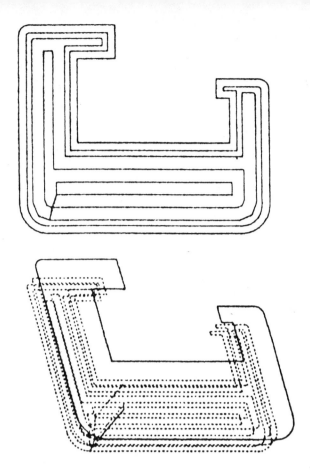

Fig. 5

Fig. 6

The AD-2000 Lathe function provides the following major functions:

    TURN, FACE, BORE,
    CONTOUR, GROOVE, TAPER,
    THREAD, TOOL

and uses the following parameters and descriptive data:

    ATANGL, BLANK, COOLANT,
    DEPTH, ENDPOINT, ENGAGE
    ANGLE, FINISH, FEED,
    HOLDBACK, LIFT ANGLE,
    RETRACT ANGLE, RETRACT
    DISTANCE, ROUGH CUTS,
    ROUGH FEED, RPM, SAFE,
    SCALLOP, SFM, SPINDLE
    DIRECTION, SPINDLE
    SPEED, START LINE,
    START POINT, STEP
    and THICKNESS

Geometry for BLANK and CONTOUR
is defined using the standard
AD-2000 construction forms.
The user is provided with
control over each operation
without sacrificing automated
sequences.

Fig. 7

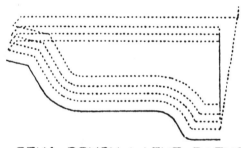

SEMI-ROUGH LATHE PATHS
(DISTANCES EXAGGERATED
FOR DISPLAY PURPOSES)

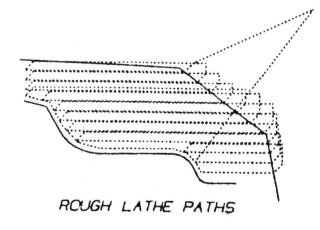

ROUGH LATHE PATHS

Fig. 8

The AD-2000 3-Axis and 5-Axis Contouring functions provide:

1. Separate rough and finish passes (with optional separate tools).
2. Cornering feedrate control is provided by a CLFILE edit feature.
3. Islands (bosses) are automatically roughed and finished.
4. A rough distance from all drive/check surfaces (including part surface boundary) can be kept. Profiling to a surface then allows a final clean-up around the edge of the machined area.

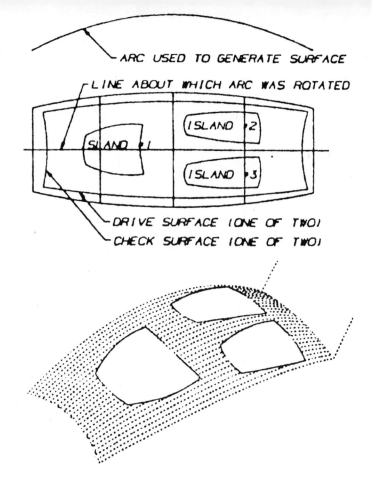

Fig. 9

Fig. 10

The AD-2000 Point-to-Point function provides:

1. Deep hole (chip relief) sequence.
2. User created drill cycle, i.e., any combination of FEED, RAPID, DWELL at either assigned depths (both negative and/or positive). Sequences are numbers and can be verified and used repeatedly for one part.
3. Thru (piercing) function which allows two surfaces to act as the TOP and BOTTOM of a drill sequence.
4. Projection of points to 3-D surface for 3-axis, 4-axis, and 5-axis point-to-point.
5. Output G80 series for applicable drill cycles for machines whose controller has automatic cycles.

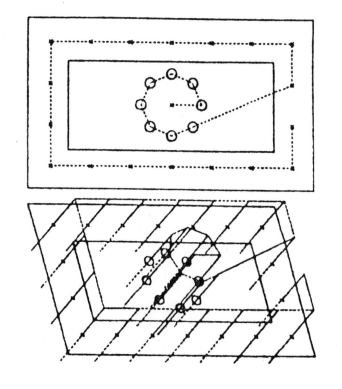

Fig. 11

Fig. 12

The AD-2000 Absolute Motion facility allows any geometric entity, or set of entities, to be used in positioning a tool absolutely, i.e., drive the tool to a specific point, through a set of three-dimensional points, or relative to any curve(s). When relative to a curve, the tool can either follow the curve or maintain tangency on either side. It can also be offset a constant distance in a principal direction.

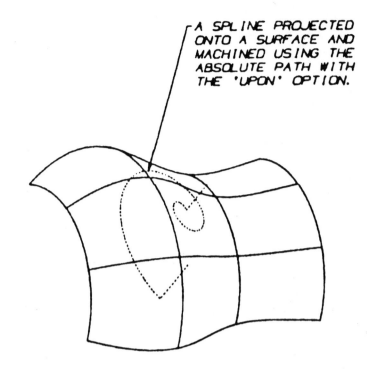

A SPLINE PROJECTED ONTO A SURFACE AND MACHINED USING THE ABSOLUTE PATH WITH THE 'UPON' OPTION.

Fig. 13

Fig. 14

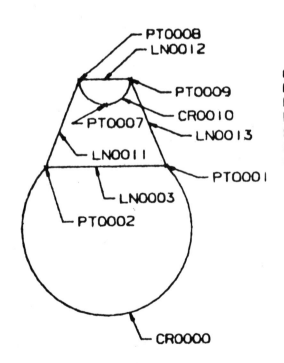

Fig. 15

```
CR0000=CIRCLE/150.00,120.00,0.00,40.00
PT0001=POINT/CR0000,ATANGL,45.00
PT0002=POINT/CR0000,ATANGL,135.00
LN0003=LINE/PT0001,PT0002
LN0004=LINE/PARLEL,LN0003,YLARGE,40.00
LN0005=LINE/PT0002,ATANGL,45.00,LN0003
LN0006=LINE/PT0001,ATANGL,135.00,LN0003
PT0007=POINT/INTOF,LN0004,LN0005
PT0008=POINT/INTOF,LN0004,LN0006
PT0009=POINT/INTOF,LN0005,LN0006
CR0010=CIRCLE/PT0008,PT0009,PT0007
LN0011=LINE/PT0002,PT0008
LN0012=LINE/PT0008,PT0007
LN0013=LINE/PT0007,PT0001
```

ATTRIBUTE MANAGEMENT
1. CREATE
• 2. INTERROGATE
3. DELETE

1ST SEARCH FORM
1. ATTRIBUTE
• 2. SUB-ATTRIBUTE
3. BOTH

SELECT MODE
• 1. RETRIEVE
2. IDENTIFY MINIMUM
3. IDENTIFY MAXIMUM
4. FIND TOTAL
5. CONSTRAINED RETRIEVE
6. DISPLAY

ENTER DESCRIPTOR
• CAM
73 CAMS FOUND.

2ND SEARCH FORM
1. ATTRIBUTE
2. SUB-ATTRIBUTE
• 3. BOTH

SELECT MODE
• 1. RETRIEVE
2. IDENTIFY MINIMUM
3. IDENTIFY MAXIMUM
4. FIND TOTAL
5. CONSTRAINED RETRIEVE
6. DISPLAY

ENTER DESCRIPTOR
• STAINLESS
42 STAINLESS FOUND.

3RD SEARCH FORM
1. ATTRIBUTE
• 2. SUB-ATTRIBUTE
3. BOTH

SELECT MODE
1. RETRIEVE
2. IDENTIFY MINIMUM
3. IDENTIFY MAXIMUM
4. FIND TOTAL
• 5. CONSTRAINED RETRIEVE
6. DISPLAY

ENTER DESCRIPTOR
• PERIMETER

CONSTRAINT RELATIONALS
1. LESS THAN
2. LESS THAN OR EQUAL
3. EQUAL
4. NOT EQUAL
5. GREATER THAN OR EQUAL
• 6. GREATER THAN

• 1. VALUE   : 450.00

17 PERIMETERS GREATER THAN 450.00
4TH SEARCH FORM
• 1. ATTRIBUTE
2. SUB-ATTRIBUTE
3. BOTH

SELECT MODE
1. RETRIEVE
2. IDENTIFY MINIMUM
3. IDENTIFY MAXIMUM
4. FIND TOTAL
5. CONSTRAINED RETRIEVE
• 6. DISPLAY

ENTER DESCRIPTOR
• VENDOR
AJAX TOOLS
M AND F TOOLING
COGSWORTH CAM LTD.
MCS, INC.
T. J. JONES LTD.

Fig. 16

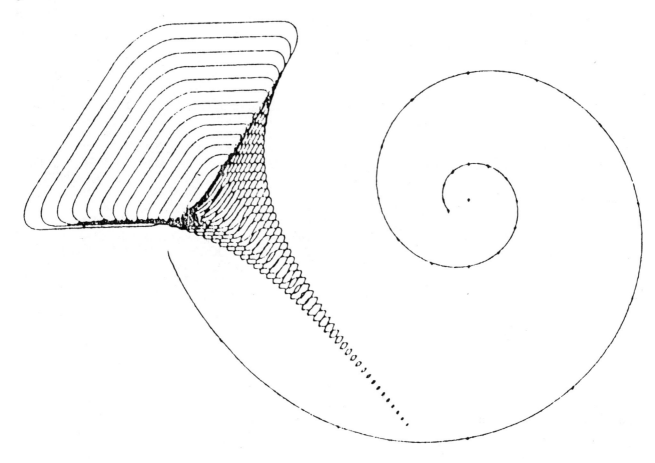

Fig. 17

Reprinted from Modern Machine Shop, January 1978

Computer Graphics Organize the Nest

The best nesting computer is the human brain. But converting the nest to an error-free numerically controlled flame-cutting program is best done with some graphics help.

By KEN GETTELMAN, Associate Editor

The torches are cutting one workpiece after another from the huge 12- by 24-foot plate. The NC program is functioning without flaw and the visitor stands in awe as the cleanly cut parts drop out, leaving only a bare minimum of web and scrap. It all looks so simple.

It wasn't always so. As recently as 1974 the plate-cutting operation at the McNally Pittsburg Manufacturing Corporation plant in Pittsburg, Kansas, followed the traditional methodology. Plates were chosen, the individual workpieces were laboriously laid out and workers crawled across the plates with hand torches to rough out the individual parts which then became components of coal handling and preparation machinery found at mines and coal-burning installations all over the world.

The comeback of coal and the rising manufacturing cost prompted Joe Keller, Vice President of Manufacturing, and Homer Livingston, then an Industrial Engineer, to tackle the problem of plate cutting to see if the bottleneck could not be broken. They were not looking for simple improvements—they were seeking a conceptual breakthrough. It came in the form of a' Messer Grieshiem NC flame-cutting machine with a huge 9.3- by 24-meter table (31 by 80 feet) with both conventional oxyfuel and plasma arc cutting heads. It was equipped with a General Automation SPC 16/45 com-

Flame cutting unit with both oxyacetylene and plasma are cutting torches. The plasma arc torches are used to cleanly cut thinner plates while the oxyacetylene torches are required for thicker plates. Part nesting is done with the aid of computer graphics and executed by NC.

puter numerical control. In June 1973, a UNIAPT programming language and minicomputer were utilized so that programming efforts could be made more productive. This period prior to startup was utilized in training of programming personnel. The flame cutter became operational in October 1974. According to

both theory and practice, all should have gone forward with minimum delay. But one problem did not yield to conventional NC discipline.

One Unique Difference

NC flame cutting differs in one very important respect from most numerical control processing. The

217

program executed on a machining center or turning center will have one unit of raw material for each workpiece, such as a casting, forging, piece of bar stock or plate. Each unit is complete in itself. On the other hand, most flame cutting involves nesting a quantity of different workpieces on a large plate in a pattern to minimize scrap. The individual parts must be programmed in terms of their locations on the plate before flame cutting can begin.

The Original Plan

The complete coal preparation systems produced by McNally Pittsburg are custom installations. A typical installation includes conveyors, hoppers, washers, crushers, dryers, and handling equipment capable of processing several thousand tons of coal per hour. Thus, each installation has many custom designed components made from low-carbon steel, high-carbon steel, alloy steel, stainless, and composite steels ranging in thickness from ⅛ to 1½ inches. Most of the components are flame cut from plate and then fabricated into large weldments. Often, the components are so large that final welding must be done at the time of installation of the coal preparation system.

Processing followed very much a traditional approach, with the design engineering department conceiving and then producing the drawings for each installation. The drawings were then released to manufacturing process planning where they were grouped according to material, schedule of when needed, time to get through the shop, and all the other usual considerations that go into scheduling workpieces for production operations.

By the very nature of the workpieces, they tended to be grouped according to material to facilitate processing by the flame-cutting method of production. When NC entered the picture, it became necessary to program a complete plate of parts. This involved selecting workpieces for a plate and then programming each in relation to the other as they would appear on the plate. In the final anal-

Programmer uses the computer graphics terminal with viewing scopes to call up and manipulate individual components. Once located on the plate the computer "translates" all workpiece data to an absolute set of coordinates for automatic generation of a flawless NC tape.

ysis, 20 or 30 different workpieces or more actually became components of a single program for cutting an entire plate.

Each workpiece could be programmed as an entity unto itself without regard to any other part that would become a component in an entire plate layout. That much presented no real problem. At the outset, the programmers were trained in the use of the UNIAPT language and had no trouble working in lines, arcs, circles, or even complex curves on a two-dimensional surface. The problem came in combining them in one complete program for processing an entire plate. The definition of a circle in relation to its center is one thing. The definition of a circle in relation to a number of components on a plate is something else. And it was here that the gritty day-to-day problems took place.

Plan and try as best they could, the programmers discovered that although they could easily define the

outline of any given workpiece, the actual translation of many workpieces to a steel plate was very difficult and time consuming. They were constantly being called to the shop because a flaw would be discovered in the NC program, which would have to be corrected. The programmers found that they were spending one hour in the shop for each hour they were writing programs. Sometimes they would work for 16 to 18 hours at a stretch untangling programming bugs as they showed up in the flame-cutting process, during the initial startup period.

The Solution

Once it was determined that "translating" individual workpieces from an outline definition to a specific relationship with many other workpieces on a plate was the real problem, the search for the solution began. NC flame cutting, even with the programming problems, was well established and had proved its worth. Thus, the question was not one of considering an overall conceptual change, but rather perfecting the structure that had been created.

The answer was found in a computer graphics system from Computervision Corporation of Bedford, Massachusetts. It consists of two terminals with CRT (cathode ray tube) scopes, a keyboard, interactive tablet, electronic pen, and access to a computer. When the computer graphics installation arrived on the scene in April 1976, no major programming or data processing procedures had to be drastically revised. The graphics system merely became the interface method of gathering data on a nest of workpieces and organizing the outline definitiion of the individual workpieces into a single flame-cutting program, utilizing an extensive program capability previously developed.

The product designers design each installation to meet specific coal handling and preparation needs. The engineering drawings come to the manufacturing processing department where they are grouped into workable lots—often five component structures. The processing department, which includes programming, assigns each individual workpiece a number and notes the type of material involved. Each workpiece is programmed in the UNIAPT language in terms of workpiece outline only. All descriptive text and fabrication instructions are made on hard copy by the graphics system. The scale drawings are made from CL data and output on the Calcomp plotter. These "paper dolls," as they are affectionately called, are cut out by programmers and filed for future reference. Based upon scheduling, workpiece material, and size, logical groups of workpieces are listed by material, type, and scheduled manufacturing start date, and output on separate lists.

The Graphics System

The first step is to call all pieces for one list up on the graphics terminal and store them on the upper part of the screen. The programmer then gets into the "translation" work with the graphics system to pin down the

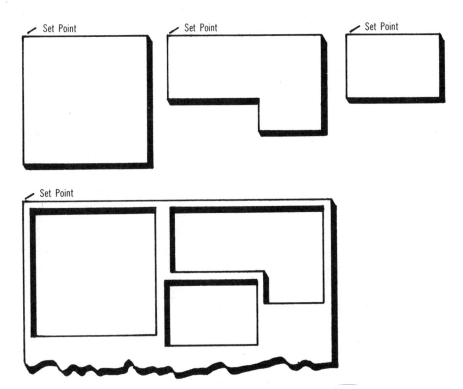

Workpieces are defined in terms of their own geometry by UNIAPT programming. Computer graphics "translate" the individual workpieces to an integrated program of many workpieces located on a single plate ready for flame cutting.

complete layout in errorless mathematical terms suitable for NC program definition. The workpieces have already been outlined in UNIAPT programming terms and the data has been stored in the computer data bank. The programmer, working with his terminal keyboard and electronic pen, selects the first workpiece to nest and with the aid of the pen moves it around on the plate diagram (which has already been outlined and placed on the viewing screen) to bring it into its exact location, from each edge of the plate. He will then call up the second workpiece and then specify the datum point of the workpiece in exact terms, by keyboard entry, to the datum point of the first workpiece. Thus, the two are absolutely located to each other on the plate—an ideal condition for creating a complete flame-cutting program.

The programmer proceeds until he has located all the workpieces. As he dimensionally describes the datum point of one workpiece to the datum point of another, the graphics

system will scale them on the viewing screen so the programmer can see if they do actually fit as he had visualized and manipulated them. As Mr. Livingston points out, "The human brain is still the best nesting computer we can find." In any nesting procedure the large rectangular workpieces are located first and others are fitted into place to form a workable nest.

As each workpiece is positively located, the geometric data is processed by the computer to position the workpiece outline in relation to the plate from which it will be flame cut. Once the plate diagram has been filled and visually approved, the programmer instructs the computer to generate a flame-cutting tape for the entire plate layout. The job may take some two or three minutes of computer time. The tape punching itself takes only a short time.

Once the tape is punched, the programmer has one more resource to ascertain the error-free nature of his program. The punched tape is run through a hard-copy plotter which

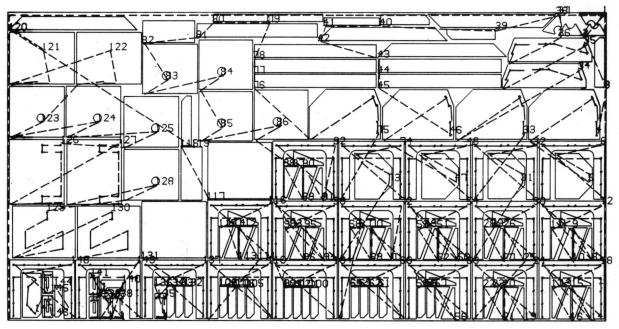

A hard-copy plotter makes a scale drawing of the actual cuts that will be made by the flame cutter. Layout lines for subsequent bending or hole drilling may also be placed on the plate by a pneumatic punch head, which is normally offset. The punch produces a series of indentations to indicate a bend line or hole to be drilled. When layout is completed, the punch head returns to its offset position and the torch heads come into play. The flame cutter spends about one-third of its time for layout and the other two-thirds for flame cutting.

Line Code	Operation
Black, Solid	Straight Cut, Single Torch
Black, Dashed	Bevel Cut, Left Torch
Black, Dotted	Bevel Cut, Right Torch
Blue, Solid	Bevel Cut, Left and Center Torches
Blue, Dashed	Rapid Traverse, No Cutting
Blue, Dotted	Bevel Cut, Right and Center Torches
Red, Solid	Bevel Cut, Left and Right Torches
Red, Dashed	Punch Head On, Rapid Traverse
Red, Dotted	Punch Head On, Operating Mode
Red, Super-Dotted	Triple Bevel, All Three Torches

makes a scale drawing of the actual cuts that will be made by the flame cutter. The various functions of the flame cutter, including completely editing all auxiliary functions and cutting modes, are verified by this software package. All machine operations are indicated by colored pens and various length dashed lines. Duration of each operation is noted and composite run time is given for a total tape. This provides a final review and also a working document for the machine operator so that he knows just how the program will function.

Results

Any change in procedure or introduction of new equipment has two distinct earning potentials—direct and indirect. The direct appeals to accountants and has the capability of showing up immediately and can be directly measured. There are many plants that make the mistake of considering only the direct savings in a question of equipment justification. This did not hold true at McNally Pittsburg. As Mr. Livingston points out, there are some very nice productivity gains, which are directly measurable. The complete NC flame cutter and computer graphics programming system increased plate production over 57 percent the first year. Toward the end of the second year it appears the productivity gains will be some 84 percent over the base period. Truly the plate-cutting bottleneck has been broken and the programmers are no longer racing to the shop for debugging.

What about the indirect? It hasn't been ignored. Scheduling was improved to the point where the increase in business did not force the construction of an additional building just to hold the various work-in-progress components. The NC flame cutter moves work through in an optimum amount of time.

The industrial engineering depart-ment has discovered that processing all work through one machine has not created a bottleneck. Quite the contrary has taken place. Running all work through a single, efficient machine has given the department much better scheduling and planning control.

As a result of easily shifting and moving workpieces around on the plate by graphics before any firm commitment is made to a final layout, there has been a 25 percent reduction in scrap. The industrial engineering department can specify the safe minimum webbing allowed between components on the plate and the computer programs to the optimum figure.

Often in the past some workpieces that should have been one component were broken up into two or three for manual processing. Then they were welded together. Now it is possible to cut much larger workpieces and even bend them after cut-

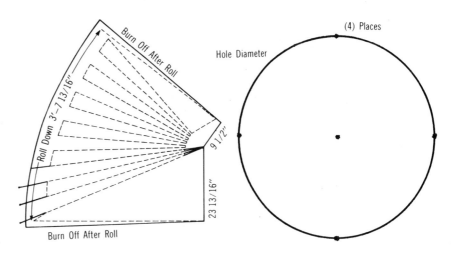

A hole layout pattern and a radial layout for a square-to-round transition pattern. Flame cutting can be very simple—such as point-to-point, straight lines and circular arcs—or highly complex, such as smooth cylindrical curves passing through a series of points, lofted conics and the development of flat patterns through the use of cubic equations.

ting. This has been a great help to product design. They can now design larger individual components, which leads to better overall product concept and performance. It also allows them to design a better product.

Finally, the system, as does all computer-based systems, has imposed a certain discipline to make it work. This has forced better procedures, the elimination of wasteful practices that tend to creep into any system that has been around for a long time, and a more logical and rational approach. Sloppy or fuzzy procedures just will not function with any kind of a computerized approach. Thus, a disciplined approach generates a more efficient operation by forcing the controlling engineering staff to think through each step of the way and to function in a more precise, disciplined manner.

Considerations

The old maxim, "There is no such thing as a free lunch" certainly applies to the achievement brought about by Mr. Livingston and Mr. Keller. Each step of the way required up to two years of good solid homework and planning. They not only had to make a choice of equipment, but they had to plan the all-impor-

tant processing and the staff to execute it.

The present equipment needed to program, nest, and prove-out the complex programs includes a Computervision DCC 116 processor, which is essentially a 64k central minicomputer; two VT-50 message terminals (keyboards); two 19-inch CRT (cathode ray tube) viewing scopes with interactive tablet and electronic pen; a Tektronix hard copy graphics unit multiplexed to the VT-50's; a 14-million-word computer storage disc; an ASR-33 teletype unit; a 45-inch-per-second magnetic tape drive; a General Automation SPC 16-65 with 24k; a 25-inch-per-second magnetic tape drive; a 400-card-per-minute card reader; a 600-line-per minute printer; a model 936 Calcomp plotting unit; two 1.25-million-word disc drives; a 12.8-million-word disc drive; a 75 cps paper tape punch; and a 400 cps paper tape reader. This equipment serves the programmers and it was purchased outright. In addition, the CNC at the machine tool has ample storage capacity at the present time to handle the volume of data needed for large plate processing. Future technology, which includes faster cutting speeds

for plasma, will require more storage. The flame-cutting machine itself is a very significant expenditure for both the machine and its installation.

A key to any operation of this type is the personnel who make it function. The programmers responsible for nesting, layout, and data handling are young mathematics graduates who are still in their 20's. They grew up with computers and have relished the challenge of making them work in the manufacturing environment. They are bright, inquisitive, and imaginative. Mr. Keller and Mr. Livingston are well experienced in the real world of manufacturing and superimpose the day-to-day "practical" knowledge so essential to a successful manufacturing function that is the result of computer processing techniques.

Although the UNIAPT program was used as the basis for workpiece programming and standard computer graphics software packages were available, much of the actual translation software had to be developed and debugged within the McNally Pittsburg facility. This approach, while more difficult, means the staff could create software to meet the company's requirement rather than make compromises to get a system on-line and functioning.

Pittsburg, Kansas, is not one of the nation's leading metropolitan areas. Computer organizations do not establish regional sales and service offices in the city. Knowing that they were on the far end of the line for service and software facilities prompted the company to make the basic decision that they would simply master both areas and handle them with their own staff. It was done at the price of obtaining adequate staffing and training; however, the approach works and that is the principal requirement.

Final Payoff

It could be said the final payoff is the virtually flawless function of the system with workpieces flowing through on schedule, no frantic calls from the shop because of programming errors, better products, and significant gains in productivity and efficiency to meet the needs of the market place. But in the final analysis, the work is never finished. New programming horizons, better programming techniques, and a wider application of the computer to the entire manufacturing operation is in the future for the dynamically growing and evolving computer-aided-manufacturing methodology.   **MMS**

Reprinted by the Society of Manufacturing Engineers from IRON AGE, June 12, 1978; Chilton Company; 1978

LATEST GRAPHICS SYSTEMS RECAST CAD/CAM CONCEPTS

By James B. Pond

Long awaited breakthroughs in interactive graphics systems offer big changes in manufacturing and engineering.

Integrated interactive design and manufacturing is the way of the future. It involves new thinking about how we run our factories. It won't come easily. But it is coming and there isn't much you can do to stop it.

Thanks to mini- and microcomputers, future technology is close at hand. Recent developments in interactive graphics systems show capabilities that go beyond what most of us even dream possible. These developments are changing notions about CAD/CAM. They are breakthroughs that are smashing down productivity barriers.

Wild rhetoric? Possibly. Everyone decries the shortage of capital to buy new capital equipment. Yet, when a really cost-effective machine, method or process is proposed, the dollars are there.

This is amply demonstrated in the jewelry-making industry which is presently clamouring for a new fully interactive system that includes machine tools as well as three-dimensional interactive graphics.

This computer-assisted design/manufacturing system completely eliminates tedious, and costly, drafting and parts-programming stages of manufacturing. In actual production it is slashing the design-to-finish mold time from 3 months to 6 hours. Conventional detailed molds can be produced, from start to finish, in less than one hour.

"A more-general average would be a 15-to-1 time-reduction ratio," says John Titus, vice president of the CDM division of National Computer Systems, Minneapolis, Minn. "But, even if the time savings were nil, our CompuTool System offers such benefits as improved dimensional accuracy, improved repeatability and improved conceptual accuracy."

The CompuTool System, at prices starting at $600,000, is so cost effective that firms with no experience with numerical control are queuing up, some with multiple orders. As one user enthusiastically exclaimed, "It is 100-years of productivity improvements in one year!"

Fortunately, CDM's concepts are not confined to the production of small jewelry items. Mr. Titus said that the CompuTool System is being offered with large machine tools such as the Hillyer machining center and a New England Machine Tool profile milling machine. He said he was negotiating with other machine tool builders in the U.S. and overseas.

The spectacular nature of CDM's CompuTool System so overshadows other significant developments in interactive CAD/CAM systems that rather than elaborate on it at this point, the CompuTool System is described separately elsewhere in this article.

There is no question that widespread interest in interactive graphics is perking up. Conceptualists view the market with considerable optimism. The market potential is huge. As a result, the number of firms seriously offering interactive graphics systems for the metalworking industry has more than doubled in the past two years.

The whole tenor of the industry is changing. Those with experience, the large corporations which bore the brunt of development headaches and costs, are beginning to explain their successes and their failures. It is a sign of growing maturity and confidence, in themselves and their expertise.

Interestingly, these are the people who are emphasizing the need for integrated interactive CAD/CAM systems. This is the reoccurring theme in many of the presentations given at recent seminars and conferences held by the NCS, SME, CAM-I, IEEE, and others.

One happy surprise in this respect is the participation of Pratt & Whitney Aircraft in various CAD/CAM conferences, not to sell a system but to alert industry of the potential benefits and pitfalls of interactive graphics and all it implies

Dr. Edwin N. Nilson, manager, Technical and Management Data Systems, P&WA, describes himself as a mathematician. He is also very practical person who understands manufacturing.

Addressing the Numerical Control Society in Chicago, Dr. Nilson put things into perspective with an overview of the P&WA CAD/CAM system. He said that it has been in development for the past ten years. It is not completed, but most of the major elements are in place and all aspects are operational in some form.

The basic structure of the system features an integrated data base in the computer which contains the description of the part (or component) and its properties. "The system is heavily analysis-oriented involving aerodynamics, structures and vibrations, heat conduction and heat transfer as well as their interactions. The use of interactive computing in design analysis constituted most of our initial entry into interactive CAD/CAM," he said.

An important aspect of this system is its capability to rapidly transmit design data between groups. It is set up so that many people in various groups can work on the same data simultaneously. The cataloging, storage and retrieval of information is automatic. Manual intervention is involved only for archival storage.

Dr. Nilson said that the benefits of the system have been substantial. "Lead times have been consistently reduced by one-half, while computing-per-design task has been reduced 30 pct. Labor-per-design task is reduced by 80 pct. Part of this advantage is used in improving designs. Generally, the productivity of the engineer is increased by at least 5-to-1 through CAD.

"These benefits in design have been matched in tool design/make and NC programming. And, in those situations in which there is a well-developed interface between CAD and CAM, the cost- and lead-time-reduction factors are phenomenal, ranging from 25-to-1 to many-hundreds-to-1."

Describing their implementation problems, Dr. Nilson said that a number of special problems arise during the early stages of implementing an integrated interactive CAD/CAM system. "Batch-type computing has generally become solidly entrenched. And, in both engineering and manufacturing, there is initially considerable reluctance to making such a fundamental change as going to integrated interactive CAD/CAM. It is, therefore essential that the man/computer interface be made as simple

and efficient as possible. The computer must provide the user with the information he needs to make his decision or judgment."

This is not easily done, he added. It is a difficult task to get the user to specify explicitly what he needs.

"The integrated data base, which now becomes the means of technical communication between groups and between design and manufacturing, is a new concept. Its proper handling requires close cooperation among engineers, applications programmers, and computer systems programmers. Indeed, close cooperation and coordination of these groups is important for the success of the program."

Dr. Nilson is adamant about the need to communicate and how it is done. "In its narrowest form, the mechanism for transferring design information to manufacturing is the drawing. In its broader form, it covers the process of interdepartmental interaction during the time that a design concept is being given precise form.

"Here, the designer should know not only what are the structural properties of a particular part but also the problems of fabrication likely to be encountered. Manufacturing people should be permitted to react during the early stages of design.

"Interactive CAD places highest priorities upon what benefits engineering most. On the other hand, the benefits to be derived from a well-integrated total CAD/CAM system are substantially larger than those derivable from either interactive CAD or interactive CAM separately. Thus, it is the integrated point of view that should prevail."

Describing their experience at P&WA, Dr. Nilson noted that work habits begin to change after the introduction of interaction to or between engineering and manufacturing. Standards imposed upon computing become more stringent. Design configuration control is achieved by new and different means. The tools provided to the engineer are quickly outgrown and new interactive tools are demanded.

"Effects that are much more subtle and harder to anticipate are the changes in the design or manufacturing engineering processes themselves which evolve as the interactive CAD/CAM system is developed. The two areas closest to the CAD/CAM interface—namely drafting and NC programming—are likely to experience the greatest changes.

"Drafting is clearly affected because the very nature of the communication to manufacturing changes. In some cases, configurations are created by designers, and even design analysts, without intervention by draftsmen. In the P&WA system, for example, all airfoil surfaces for fan, compressors and turbines are developed in design analysis and neither mechanical design nor drafting contributes to their shape. Engineering master drawings, where required, are produced automatically by the computer.

"On the manufacturing side, NC machining represents the oldest of the technical applications of large-scale computing. And, just as there is reluctance in design to switch from batch computing to interactive computing and then to integrated interactive computing, so is there reluctance in NC programming. It is solidly entrenched in a batch mode. And the pressure to produce error-free NC programs militates against any change from the tried-and-true methods.

"On the other hand, NC programming is probably the area which can best capitalize upon direct access to part description in the design data base in the computer. Here we see 2-to-1 reduction in lead time and 5 or 6-to-1 reduction in labor being borne out in case after case, with far greater savings frequently attained.

"In numerical control we shall probably see the most radical changes in the manner in which programming is carried out. For example, at P&WA we have recently completed the construction of an interactive computer system. In this system, the NC programmer sits at a scope terminal and graphically lays out his cutter path. The APT programming and post processing are carried out automatically as is the punching of the control tape for the machine. There are no drawings.

"As the various machines are included in a direct numerical control (DNC) process, of course, the tape will be eliminated. Note that the use of APT here is a temporary expedient, held in place by the existing form of the post processor. A much less tenuous link between scope and machine tool is inevitable. The dependence upon particular NC programming language such as APT, together with the associated post-processors, will be substantially reduced and even eliminated from the scope/machine tool relationship. The

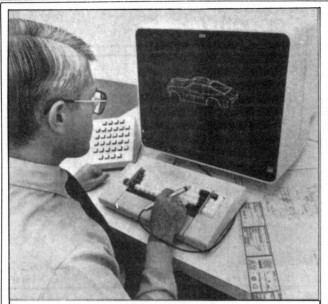

IBM's 3250 graphics display system allows operator to push a key to generate standardized symbols.

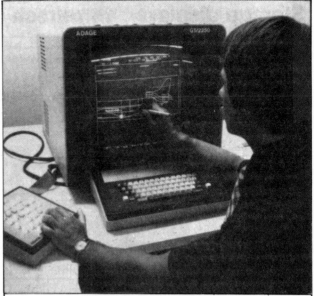

This display of an aircraft on an Adage CRT gives only a hint of the complexity of design capability.

complete task is then essentially converted to a process planning function."

Dr. Nilson feels that CAD/CAM is advancing at a significant rate and that any CAD/CAM system should be flexible and capable of modification and growth. It should also be highly productive at any point in time.

For this reason, he cautions about complete reliance upon the vendor for the development and incorporation of changes. "This leaves the user vulnerable," he says, "and it will delay significantly the evolution of CAD/CAM in a given installation. The required special expertise is in short supply, even with vendors. What makes matters worse, vendors have to concentrate on meeting the most pressing demands of the majority of their users. This can result in special situations going unattended for years."

To make matters worse, he says, minicomputer-based interactive drafting systems are frequently constructed with machine systems which are not easily accessible to the user. And size limitations, together with problems associated with integrating an interactive drafting system into a large integrated CAD/CAM system, create demands for special computing systems expertise on the part of the user.

Getting started in interactive CAD/CAM via the route which concentrates on interactive drafting plus NC programming has its pitfalls, he warns. "Though significant rewards will result, this initial surge into the new CAD/CAM technology may well result in both interactive drafting and interactive NC programming as

such becoming solidly entrenched. This would result in postponing or even delaying indefinitely the evolutionary changes which should take place. This would make the larger integration into a total integrated interactive CAD/CAM system harder to attain."

Speaking out on the effects of geometric modeling and group technology technologies, Dr. Nilson sees no serious problems in the implementation of their CAD/CAM system. "One characteristic of many design/manufacturing processes ignored or overlooked by those concerned with advancing these technologies is stability: stability in the design process, and stability in the process-planning process.

"What we mean by stability in design is that in moving from one design to another, much of the design process is unchanged. Changes can always be handled on an exception basis. Similarly for process planning. At P&WA we have been able to capitalize upon such stability in our interactive CAD/CAM structure.

"One area where advancing technology has already begun to pay off is in design/manufacturing communication. I refer to the process in which a three-dimensional concept in the mind of a designer is represented by traditional two-dimensional drawings via orthographic and isometric projections, communicated in this form to a process planner or NC programmer, and then reconstituted as a three-dimensional concept in the mind of the latter.

"Three-dimensional graphics is an effective answer to this problem. The process planner can rotate the three-

dimensional object in three dimensions on the scope and immediately get the 3-D concept. Only with 3-D graphics does the designer get proper feedback which tells whether or not the two-dimensional drawings he has made properly represent his concept."

The fact that CAD/CAM technology is in a state of change makes it difficult to comment on what CAD/CAM systems are available at any one moment.

"The CADAM system of Lockheed's, which is now being distributed by IBM, is primarily a drafting system with NC capability. CADAM runs in IBM 360 or 370 hardware with time-sharing. It is at the larger-computer end of these product lines. It contains an analysis package which permits a certain amount of structural analysis.

"The CADD and CADM systems of McDonnell Douglas constitute a roughly comparable system. Each of these also runs on IBM hardware. Very heavy support of forty IBM 2250 light-pen scopes at McDonnell Aviation is provided by six 370/168's with up to 10 megabytes of core memory. Again, structural analysis and NC programming capability are included.

"Generally, these systems contain, for software, a design/drafting package, a geometry module, an NC programming package, some elements of structural and vibration analysis, and a data base capability with its storage and retrieval functions," he explained.

Design/drafting systems built around minicomputers are offered by Applicon, Auto-trol, Calma, Computervision, Gerber Scientific, Information Displays, and United Computing. Most provide NC programming capability. Links of these small interactive drafting systems to large computers are available in some cases. These links are essential in the integration of total CAD/CAM systems. There are also design/drafting software packages (software) available, such as the AD-2000 System from MCS, that can be installed in a variety of computers.

"There is only system like P&WA's which spans the entire range of technical CAD/CAM (as opposed to management information systems support) in design and manufacturing. That one is IPAD (Integrated Programs for Aerospace vehicle Design), which is not commercially available at this time. IPAD is being developed by Boeing Commercial Airplane Co. under a five-year NASA contract ending in 1981 (first phase).

"In short, we find large and small systems, off-the-shelf and home-grown systems, with wide variation in system design philosophy. Part of this wide variation is due to the fact that CAD/CAM technology is in the process of evolution, part is due to basic differences which result between manufacturing and engineering when a common meeting ground is being established, and part is due to the wide range of industrial settings which such systems are to support.

The smaller minicomputer-based interactive drafting systems, Dr. Nilson noted, are just that: drafting systems. For some users, nothing more is needed, now or in the foreseeable future.

Although Dr. Nilson's comments on minicomputer-based systems may indicate a bias to large computer-based systems, he said that P&WA has purchased a large turnkey minicomputer-based system recently.

Robert A. Tomisak, engineering design supervisor, 3M Co., Minneapolis, Minn., has also given a paper answering questions asked about CAD/CAM and interactive graphics. He spoke at the SME's Autofact II.

Regarding justification of these systems, he noted that they are costly. Prices for a basic one-terminal system, complete with hardware and software, start at well over $100,000.

They can be made more affordable, he explained, because most popular minicomputer-based systems are capable of supporting multiple CRT terminals. Where this is possible, unit terminal cost may be reduced to something like $50,000. Another cost-cutting step is to use a reel-to-reel tape drive to interface the CAD system computer with existing plotters.

CompuTool eliminates drafting and NC programming

One of the most dramatic breakthroughs on the CAD/CAM and numerical control scene today is the CompuTool System developed by the CDM division of National Computer Systems, Minneapolis.

It is a fully-integrated system, complete with refresh 3D interactive graphics and appropriate machine tools.

The system, described at the 1975 annual technical conference of the Numerical Control Society held in Washington, D.C., did not have a significant impact at the time, apparently because it had a very specific use in jewelry making.

Since that time, John Titus, vice president of CDM, reports that, in addition to tool design and tool making at jewelry firms, the system is applicable to large workpieces and machine tools. Several systems are now in the field, and a healthy backlog of orders, many of which are repeats, give added credence to this evolving technology.

Mr. Titus notes that part of CDM's success has been due to evaluating CAD/CAM from machine tools and servo-motors rather than from the computer graphics.

CompuTool consists of three subsystems: CompuScanner, CompuGraphics and a CompuGraver or CompuMill machining systems. The amazing aspect of the system is, that it can eliminate mechanical drafting and NC programming.

CompuScanner is a data-translation system that creates a general mathematical description of the product to establish overall dimensional parameters. It uses a three-dimensional measuring machine to establish X, Y and Z coordinates that are put on a magnetic tape.

The CompuScanner operates from drawings, templates or a rough model which does not have to be built to any specific scale.

The CompuGraphics system is a computer-refreshed CRT display that dynamically shows the design in three dimensions. Mr. Titus explains, that the intensity of the image on the CRT screen is brightest for items in the foreground. This **gives the operator continuous orientation or 3D perspective.**

The CRT screen allows virtually unlimited data manipulation. A design can be enlarged, reduced, rotated in any direction, dimension, in addition to being viewed from different angles.

"The designer can scale up a portion of the image, or reduce it, to get a better perspective. Using a light pen, he can smooth out coordinates, or soften tight or hard-to-work curves and angles," Mr. Titus said.

Class Ring Design

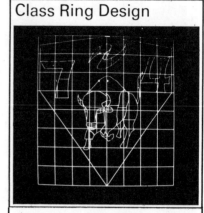

Buffalo's image, shown on CRT, can be directly machined in 3D.

"Also, predetermined manufacturing tolerances can be plugged into the final design.

"The system can pinpoint potential molding and assembly problems before they become assemblyline headaches," he said. "It can show constraint or collision between moving components or mating parts.

"The fact is that anything you see on the CompuTool CRT you can machine," Mr. Titus stated. "And, you don't have to know anything about numerical control or NC programming to machine it!"

Another intriguing aspect of the CompuTool System is the use of compact reel-to-reel magnetic tape cartridges made by 3M Co. One magnetic tape cartridge is equal to five sloppy discs at the cost of a single disc, he said. It is the equivalent of three miles of conventional NC paper tape. One of the present users requires that much capacity per day operating three shifts.

CompuMills are numerically controlled milling machines designed and built to work with the entire CompuTool System. They use data on the magnetic cartridge to simultaneously control all three axes of milling. The result is a clean mold or prototypes with straight lines, arcs and contours accurate to within 0.0001 in.

The firm has built the CompuGraver for smaller work. To handle delicate end mills, routers and engraving tools, the machine is fitted with air bearings.

Mr. Titus reveals that the CDM system monitors 27 machine tool functions and, should something fail, a message appears on the CRT.

tape punches, post-processors, other computers, etc.

Mr. Tomisak looks upon sales literature claims as being optimistic and sometimes fictional, especially when highs and lows in system productivity are averaged over long periods of use.

He notes that productivity ratios of from 1.5-to-1 to 2-to-1 will produce a return-on-investment in excess of 30 pct, which is adequate for approval in most cases.

The typical engineering organization can apply computers and interactive graphics systems to engineering analysis and design, mechanical design, and drafting.

Mr. Tomisak said that computer graphics systems are beginning to impact the engineering analysis and design area as software becomes available to emulate remote batch terminals. A typical application might involve using the local graphics system to create a finite-element model of some component. The data is then transmitted to a large computer where a program performs the analysis. The results are then transmitted back to the graphics system for postprocessing.

Mechanical design includes traditional work done of drawing boards. As examples, he cites tool design, machine design and product design.

Mr. Tomisak said the most widespread use of interactive computer graphics systems lies in drafting. "Computers and computer graphics notwithstanding, the engineering drawing will be with us into the indefinite future. It is needed to document design and as a communications aid. Drawings will be needed by inspectors, assemblers, maintenance personnel, millwrights, and others."

Computer-based design data creates a natural interface to NC manufacturing processes, he said. "This data base integration of engineering with design, and design with manufacturing, can significantly impact a large part of the lengthy cycle from the conception of an idea to the entry of a new product into the market place."

Until the past year or so, computer graphics systems fell into two broad categories: the very powerful design systems that require large computers, and the powerful minicomputer-based drafting systems. In the former case, he said, drafting is not cost-effective due to high computer costs. In the latter case, design is not cost-effective due to lack of capabilities.

This has now changed, he said.

"The better systems now provide a general purpose, high-level programming language which permit a non-professional programmer to do family-of-parts programming, create mathematical and analytical software, write computer-aided design programs, etc. Such languages take the best of FORTRAN, APL, APT, BASIC and SNOBOL to create a total graphics capability.

"The ability of a system to produce graphics from user-supplied parameters can be put to use in a second important design area. Noncreative labor spent in drawing frequently used or repetitive components or arrays can be eliminated. Libraries of typically used items can be built up gradually and can be supplemented by exchanges with other users through user groups.

"Without this user-programming feature, an otherwise good graphics system is likely to remain strictly a drafting tool. With it, the preparation of drawings becomes a complementary part of the total engineering and design process.

"The real benefits of this technology are obtained by integrating design and drafting."

Regarding graphics NC programming, Mr. Tomisak said that most suppliers of mechanical design graphics systems also have software packages to produce NC machine tool programs. Tapes or programs for piece parts already in a design/drafting data base can be created without having to back-define geometry to a computer. Output from the graphics system may be either APT-source, CL-source, or object tapes if postprocessors reside on the graphics system.

While this appears to represent an optimal CAD/CAM interface, experience has shown that no one method exists to bridge CAD and CAM operations which is best in all cases. The single most important factor to consider is the nature of the work itself.

"A company which has developed an extensive library of macros to do family-of-parts programming that deals with complex geometry and perhaps three-axis work would find interactive graphics offering a clear advantage.

"Systems are available which run both batch and interactive NC programming systems concurrently on one minicomputer. They may represent a way to acquire both capabilities at least cost," Mr. Tomisak said.

In the final analysis, Mr. Tomisak said, systems are best selected on the basis of performance, cost and productivity. He cited graphical editing capability and characteristics as a measure of productivity. By this, he referred to the type of CRT used and related hardware and software. He also expressed interest in associative data bases which were only beginning to appear in mid-1977.

From a performance standpoint, he feels that graphics systems must be highly interactive to gain broad-based acceptance in the typical engineering organization. Running three, four or more CRT terminals on a single minicomputer degradates performance, sometimes to the point where performance is only marginally acceptable.

"High-performance one- or two-terminal demonstrations involving small graphics data bases does not necessarily imply that the same high performance with multiple terminals will exist in a real work environment," he said.

Though the cost of computers is coming down, complexity along with performance and capability have gone up. Systems cost is expected to remain at a high level. Until the price gets down to about $25,000 per CRT device or terminal, Mr. Tomisak feels the typical engineering organization will likely remain a one- or two-graphics system user.

Objections to hardware-dependent interactive graphics systems are being met by the AD-2000 system developed by Patrick Hanratty, president, Manufacturing & Consulting Services, Inc., Costa Mesa, Cal. It provides a totally computer-independent data base.

"The AD-2000 was designed and evolved from interaction with users to provide a common thread for the design, drafting, management information, and manufacturing processes," Dr. Hanratty said at the 1978 annual technical conference of the Numerical Control Society.

"I think we all know what we mean when we say manufacturing and fabrication, but there is a bit of hesitancy about what is design.

"Design, really, is the recognition that there is a need to communicate. The fulfillment of that need is the completion of the design process."

Dr. Hanratty explained that in design we have a set of real world entities and a set of definitional phenomena or natural laws which project into a graphic or design plane where the geometric model is married with the math model. This allows the interaction that takes place in going back and forth in the design process. As the designer evaluates he keeps interating the design of the math model until a finalized design is achieved.

To go through a process like that it is necessary to have a very strong and complete system which combines interactive communicating with the computer and also a total geometric data base on which to apply the math model.

Dr. Hanratty said that in establishing goals for the AD-2000 system he recognized that in today's hardware technology there is no way that you can take a system and put it on one computer or even one type of computer or family of computers and say that this is going to supply your needs forever more. Therefore, he sought portability and input/output independence.

"Once you put information into a computer, you should never have to put it in again," he said. "If you use a computer for design, that information should be available for management, to determine how long it took to design a component, what it weighs, the costs involved, etc. Information should be available for manufacturing, to make numerical control tapes, etc."

"We had to have a system that would be extensible; one whereby we could add new applications, new entities and new forms for assisting entities. We had to be able to handle new ways of defining lines and circles, surfaces, and ways of interfacing with other applications.

"We had to consider that companies that have been involved with computers for the last several years and decades have built up libraries of tools which they don't want to obsolete. There are a million packages that already exist with which you have to interface," he said.

Dr. Hanratty gives some pretty good advice. For example, he told his NCS audience that "If you don't have a complete geometric model, you can forget about manufacturing. If you have a 2-D world and you try to do 3-D machining, it won't work; you do not have the information there that is required."

In later conversation, Dr. Hanratty said that the AD-2000 system would incorporate or be capable of handling both the PADL and SynthaVision geometric modeling technologies.

MCS does not plan to market the AD-2000 system directly. Several large computer-based companies have expressed interest in marketing it. However, at press time pending contracts had not been signed.

Interactive graphics is changing

manufacturing methods. For some, it is giving management, design engineering and manufacturing people greater powers and speed of execution than was ever before possible.

As more interactive graphics systems are installed in the field, the advantages of computer-generated drafting and design will become recognized and accepted.

Interactive graphics will join the computer and numerical control as a significant benchmark in the evolution of industry.

Future systems, some soon to be announced, will fan the flames of enthusiasm, and curiosity. Actually, these new systems may slow implementation of computer graphics. It will take time to assess them. But, they go beyond automated design and drafting to encompass geometric modeling, group technology and advanced numerical control.

It will be up to the individual systems manufacturers to tell their story about their individual approaches to computer-aided design and computer-aided manufacturing. It is a difficult task. It is not a fast sell. Because of the importance of the message, however, it is to industry's advantage to open doors and minds, and listen until the message is clear and the goals understood.

Storage displays . . . refresh . . . or both?

Computer graphics is a technique used to manipulate and display data in a computer. It is an aid to understanding data that for years had been presented to users in "tab listings" form.

Because computer graphics techniques involve different hardware and software, the computer graphics beginner is easily confused by

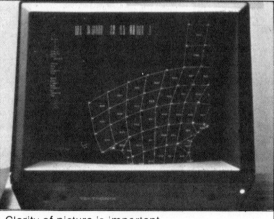

Clarity of picture is important.

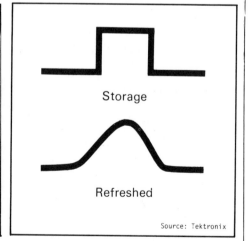

Storage

Refreshed

Source: Tektronix

the terminology bandied about in literature, seminars, demonstrations and by salesmen.

"There is only one clear-cut conclusion which can be reached about graphic displays: no one display technology is superior across all application environments," says Peter G. Cook, manager, business analysis, In-

formation Display Group, Tektronix, Inc., Beaverton, Ore.

Mr. Cook explains that there are four primary application areas, each of which having different requirements for data manipulation and display.

"The first is the presentation of data to aid human understanding. In-

cluded in these presentations are bar graphs, cartesian and logarithmic XY plots.

"The second is line drawings. Schematic and mechanical drawings, maps and integrated circuit masks are good examples. Here, the picture is the key element; the data structure is of secondary importance.

"The third involves continuous-tone pictures. These pictures consist of dots, or picture elements, called 'pixels.' Each pixel has a location, intensity and color. A commercial TV picture contains about 300,000 pixels.

"The fourth application involves the combination of graphic data with text to generate a document ready to print. The printing and publishing industry uses this technique quite substantially," Mr. Cook explains.

Mr. Cook notes that there are four display technologies of primary interest to computer graphics.

"I put calligraphic displays first on the list because these displays, which also are called dynamic graphic systems, vector refreshed displays and directed beam displays, use a cathode ray tube (CRT) similar in design to the familiar TV tube. The CRT, however, has a faster deflection system and a precision gun for more precise control of the electron beam.

"The display controller accesses a display list, or set of commands in memory, and generates the analog voltages needed to drive the deflection circuits of the display.

"The phosphor on the screen lights momentarily when the electron beam hits it. In order to keep the image visible on the screen, it is necessary to refresh, or rewrite, the image at a rate faster than the 'flicker fusion rate of the eye.' This avoids flickering of the image.

"Because the picture is redrawn 30 or more times a second, the screen image can be changed on the fly, creating a dynamic display that can interact quickly with the operator and can display picture animation.

"The second display technology is the raster scan, also called digital television (DTV). These displays operate by modulating the beam intensity of a CRT while the deflection circuits move the beam position in a regular pattern. The most prevalent use of this technology is in alphanumeric display terminals.

"The third technology is the plasma panel display based on gas discharge. This is a thin display having a good brightness and contrast ratio which has many good applications but many limitations as well. It is not used in computer interactive graphics and is mentioned here as one of the techniques being used today.

"The fourth is the direct view bi-stable storage tube display (DVST). In the DVST, the storage and maintenance of the image are inherent in the CRT display itself. The storage tube uses secondary electron emission characteristics of the phosphor to keep individual phosphor particles illuminated with low-energy flood guns once they have been activated by a high-energy writing beam.

"The storage tube technology has an inherent characteristic of very high edge acuity. In a refreshed display, the edges of the trace trail off in brightness, whereas the phosphor particles at the edge of a DVST trace are either on or off.

"The DVST phosphor reactivation is continuous because the flood of electrons create a steady DC current as they flow to the screen.

"In refresh systems the beam sweeps by at a rate of 30 or more times a second so that light emission of each point pulsates. This pulsation can lead to eye fatigue.

"These three DVST characteristics . . . continuous lines, edge acuity and continuous emission . . . combine to create a display with outstanding viewability.

"Recent developments in intelligent DVST terminals and interactive workstations have clearly demonstrated the value of computer power dedicated to DVST displays. One of the exciting developments is the combined stored and refresh display.

"The storage tube also has a capability of 'write-through' mode. The beam intensity is kept low enough to create a line but not high enough to store. The result is a refreshed image, readily distinguishable from the stored image which provides the same dynamic capabilities as a refreshed display.

"The refreshed image and the stored image appear together, providing a capability to build up an extremely complex stored picture with individual elements manipulated dynamically to the correct size and location and then stored.

"This two-in-one technique combines the advantages of stored and refreshed displays.

Effective CAM, DNC, CNC and NC System Maintenance

By Frank T. Cameron
Brown & Sharpe Manufacturing Company

Return-on-Investment is widely used to measure management performance particularly on large investments such as CAM, DNC, CNC and N/C system installations. In his new book, "Management Tasks-Responsibilities-Practices", Peter Drucker states as follows:

"Effectiveness is the foundation of success - efficiency is the minimum condition for survival after success has been achieved. Efficiency is concerned with doing things right. Effectiveness is doing the right things."

An effective maintenance program efficiently operated is one of the cornerstones required to achieve your ROI objectives. Some of the key guidelines for effective maintenance include -

1. Utilization of available scheduled hours (Direct Workers) - 85% or more.

2. Allowable N/C System Downtime for Maintenance - 8% - 10% of available scheduled hours.

3. A good P/M Program. (Preventive Maintenance)

4. Good Lubrication and Filtration of Oils to 3 Microns.

5. Weekly reporting of performance.

The real cost of your CNC system downtime is the dollar value of shipments missed - probably $75/hour or more depending on the system installed. This perspective is essential to the development and implementation of an effective maintenance program for your N/C, CNC, DNC or CAM system.

INTRODUCTION

In his recent book, "Management Tasks-Responsibilities-Practices", Peter Drucker states as follows (Chapter 4, page 45):

"Effectiveness is the foundation for success - efficiency is the minimum condition for survival after success has been achieved. Efficiency is concerned with doing things right. Effectiveness is doing the right things."

(Underlining is mine)

Mr. Drucker's comments about effectiveness and efficiency go right to the heart of our maintenance challenge for any CAM, DNC, CNC or N/C system.

GUIDELINES FOR EFFICIENT MANUFACTURING PERFORMANCE

To build an effective maintenance program, you must first know the performance standards against which manufacturing performance is to be measured. Though each situation is unique, I should like to present the following parameters for your consideration. We shall then be on the same track for the balance of this presentation.

Performance vs. Standard (Direct Workers) 95% - 100% using good measured day work standards.

Utilization of available scheduled hours (Direct Workers) 85% or more.

Allowable Machine Downtime for Maintenance -

6% - 8% of available scheduled hours for single purpose N/C machines.

8% - 10% of available scheduled hours for machining centers and more complex systems.

(These downtime values include preventive maintenance, but exclude scheduled major machine overhauls.)

Please bear in mind with CAM, DNC and CNC systems, metalworking companies are rapidly moving from a labor intensive to a capital intensive situation. Capital generates ROI (Return-on-Investment) only while being utilized.

With this introduction let's consider some of the specifics of an effective maintenance program.

ELEMENTS OF AN EFFECTIVE MAINTENANCE PROGRAM

1. Management Awareness from Top to Bottom of the Need for Effective Maintenance

    (From CEO to Front-Line Supervision)

2. Know the Real Cost per Hour of Maintenance Downtime

3. Buy the Right CAM, DNC, CNC or N/C System for the Work to be Performed

4. Make a Proper Installation -

    (1) A good foundation.

    (2) Adequate and clean electrical service.

    (3) Adequate grounding.

5. Maintenance Organization Structure -

    Separate Planning from Doing.

6. Maintenance People -

    (1) Select the best.

    (2) Train them well.

    (3) Pay them well and keep them.
    (N/C technicians should be among the highest paid people in your shop)

7. Equipment and Supplies -

    (1) Your people need the right tools.

    (2) Establish a Maintenance Stockroom.

    (3) Provide adequate maintenance spares.

8. Provide a Maintenance Work Order System

9. Effective Lubrication and Regular Filtration of Oils is a Must!

10. <u>Preventive Maintenance Program</u> -

    (1) Develop one.

    (2) Implement it.

    (3) Modify it to achieve maintenance downtime objective.

11. <u>Report Performance Weekly</u> - Keep Score!

    (1) Ratio - Hours Run to Scheduled Hours.

    (2) Ratio - Maintenance Downtime Hours to Scheduled Hours.

    (3) Keep your people informed.

12. <u>Have a Periodic Equipment or System Rebuilding Program</u>

TOWARD AN EFFECTIVE MAINTENANCE PROGRAM

1. <u>Management Awareness from Top to Bottom of the Need for Effective Maintenance</u>.

Is maintenance just a "<u>necessary evil</u>" in the eyes of your Top Management?

Does your <u>Production Management</u> say, "We'll let you work on the <u>machine or system</u> - when and if we can spare it from production"?

Do operators say, "When the machine or system goes down, we'll holler <u>long and loud</u> until Maintenance gets it fixed"?

<u>Stop right here</u>! If this sounds all too familiar, your maintenance program is <u>most likely ineffective</u>.

The record will probably **show** -

(1) <u>Your maintenance downtime is high</u>.

(2) <u>Your productivity is below par</u>.

(3) <u>Your profit rate and ROI are below your objectives</u>.

To start changing attitudes, try three places –

(1) Your tape readers –

Clean and service them weekly – a must!

(Tape readers have been the source of much maintenance downtime)

(2) Your tapes –

Are they correct?

Maintenance people have spent much time only to discover a defective tape.

Try a test tape or on a CNC system check your input and output signals.

Have your programmer re-check his tape.

(3) Filter your hydraulic oil to 3 microns each month –

Dirty hydraulic oil causes endless wear and serious maintenance downtime.

Clean hydraulic oil will last 5 to 10 years or more without a change. Both Brown & Sharpe and the Los Angeles Division of Rockwell have proved this. You can, too!

The skeptical diehards will sit up and take notice as your maintenance downtime decreases and productivity improves.

2. Know the Real Cost per Hour of Maintenance Downtime.

Your decision here is the key to the success or failure of your maintenance program!

(1) Is it $9.00 per hour – the approximate cost of your direct labor plus fringe benefits for 1978?

(2) Is it $9.00 per hour plus factory overhead?

(3) Or is it the sales value per hour of goods not produced – say $60.00 per hour or more – maybe several hundred dollars per hour for a DNC system?

In considering the impact of maintenance downtime, take into account the following factors --

(1) The more heavily committed your shop is to N/C, CNC, DNC and/or CAM, the less opportunity you have to produce parts by conventional methods.

(2) Working overtime to make up for lost time is possible - if you are not already on an overtime schedule.

(3) Subcontracting of your production is possible - provided that other shops are not as busy as you are. Lead time is also a factor.

(4) If your equipment and work are unique, outside sources may be hard to find.

(5) If unplanned downtime results in a missed shipment and a lost customer, what does that cost you?

Should you conclude that your maintenance downtime cost is $100 per hour or more, your maintenance program will be far different than one based on $9.00 per hour.

This cost of maintenance downtime is a critical factor in building an effective maintenance program.

3. Buy the Right CAM, DNC, DNC or N/C System for the Work to be Performed

   The correct system properly utilized means better productivity including better productive up-time.

   (1) Give your maintenance people an opportunity to review the system you plan to buy.

   (2) Locate other users and review their results with systems similar to the one you propose to buy.

   (3) Consider the "Task Force" organization in purchasing and installing your new system. Include a maintenance person.

   (4) Determine and evaluate the reliability in service of each major element and function of the system such as the control, tool changer, pallet shuttle positioning, machining functions, etc.

   (5) Clarify your particular requirements, and include them in your original purchase order.

4. Make a Proper Installation

Do it right - your troubles will be reduced!

Do it wrong - your troubles will multiply at an exponential rate!

Some Do's and Don'ts

(1) Your foundation should weigh at least 1 1/2 times the weight of your machine.

Most machine tool builders are skimpy on foundation depths.

(2) Isolate your foundation from the factory floor with up to 2 inches of styrofoam or other insulating material.

(3) Be aware of special vibration problems such as presses, railroads, heavy trucks, etc.

(4) Be sure your power supply to your machine and control are "clean". Don't connect to the same bus duct that services some welding equipment.

(5) Insure that you have a good ground for each control. Some builders require 5 ohms or less. Many PC Assemblies in controls cost $1,000 each. They are easily destroyed.

(6) Insure that your foundation is strong enough to keep the machine straight and level.

5. Maintenance Organization Structure -

Separate Planning from Doing.

Under a Maintenance Services Engineer, establish the following:

(1) A Maintenance Work Order System.

(2) A Preventive Maintenance Program.

(3) A Lubrication and Oil Filtration Program.

(4) A Maintenance Storeroom.

6. Maintenance People -

   You can afford the best!

   (1) Start with good supervision.

   (2) Select electronic technicians with good elec-
       tronic backgrounds - the service technical
       schools, or electronic industries. (Your
       longterm maintenance electricians usually don't
       qualify.)

   (3) Your electronic technicians should be among
       the highest paid people in your shop.

   (4) Machine tool builders and control builders
       offer excellent training programs. If you
       have a number of people to train, consider
       having the instructors come to your plant
       to conduct the training. It's cheaper that
       way - more effective, too!

   (5) Let every technician - both electronic and
       mechanical - have his own manuals.

   (6) Provide good leadership to these skilled
       technicians.

   (7) Help them to save steps. Studies show that
       up to 50% of the time of maintenance people
       is spent walking.

7. Equipment and Supplies

   To be effective and efficient, good technicians need
   proper equipment and supplies.

   (1) Establish a maintenance storeroom and stock it
       adequately. Be guided by the real cost of down-
       time.

   (2) Use a maintenance clerk to establish and maintain
       your stock records.

   (3) You will need good oscilloscopes.

   (4) Each technician should have his own portable cart
       of tools.

   (5) Provide adequate alignment tools to verify system
       component alignments.

(6) Have manuals for each technician.

(7) Telephone the Service Manager of your machine tool vendor to ask questions and confirm your findings.

(8) Learn how to benefit from Emery Air Freight - VIP Service.

8. Provide a Maintenance Work Order System

To schedule, dispatch and follow-up maintenance work, a maintenance work order system is a must.

(1) Design your own form. There are many examples available.

(2) All maintenance work should be performed from a work order.

(3) Schedule, dispatch and follow-up on preventive maintenance work using the maintenance work order form. P/M work orders can be pre-typed for scheduled dispatching.

(4) Schedule and dispatch maintenance work using a Schedule-Dispatch Board.

9. Effective Lubrication and Regular Filtration of Oils is a Must!

Nobody needs to tell us how important good lubrication is. The real question is, "How do we insure achieving good lubrication?"

(1) With the help of your lubricant supplier, insure that you are using the proper lubricants.

(2) Establish daily "oil-routes". Oilers should be equipped with lubricant carts to enable them to lubricate machines on their routes.

(3) Filter hydraulic and lubricating oils monthly down to 3 microns. Your oils should last for ten years! Use your maintenance work order form to schedule this work.

(4) Clean oils and clean hydraulic systems will eliminate much downtime and replacement of expensive components.

10. A Preventive Maintenance Program

To achieve 85% productive up-time or better you must have an **effective preventive maintenance program**.

(1) Develop your own.

(2) Use the maintenance work order form.

(3) Schedule the required work weekly, monthly, quarterly, etc.

(4) Start your **P/M Program** using information in the maintenance manuals from your system supplier.

(5) Your **P/M Program** should include -

   a. Weekly or semi-weekly air filter changes.

   b. Lubrication and oil filtration to 3 microns.

   c. Weekly cleaning of tape readers.

   d. Checking all wipers.

   e. Clean electrical and electronic cabinets.

   f. Modify your P/M Program based on experience.

Without a good P/M Program, your maintenance downtime will promptly double or triple!

11. **Report Performance Weekly** - Keep Score!

People like to know how they are doing. Two important measures are -

Per Cent Productive Up-Time to Total Scheduled Hours (should be 85% or more)

Per Cent Maintenance Downtime to Total Scheduled Hours (you can expect 6% - 10%)

(1) Publish the results weekly.

(2) Be sure your maintenance supervisors and technicians know what the results are.

(3) To achieve your ROI objectives you should equal or better the suggested targets.

(4) If you are consistently below target, find out why.

    a. <u>Correct the problems</u>.

    b. <u>Tighten up your P/M Program</u>.

    c. Your results should improve.

12. <u>Have a Periodic Equipment or System Rebuilding Program</u>

N/C systems work much harder than conventional machines - usually on a three shift basis. At the end of a period such as five years, your system will show signs of wear and deterioration.

Look for -

(1) Misalignment.

(2) Fishtailing of columns and saddles.

(3) Expended wipers - not wiping.

(4) Control circuit voltages near or below low limits.

(5) Dirt in control and electrical cabinets.

(6) Sluggish solenoid valves.

A thorough diagnostic check and major overhaul of this type will assure your continuing ability to meet your ROI objectives.

CONCLUSION

To compete effectively, you must -

1. Introduce and use efficiently the appropriate N/C, CNC, DNC or CAM system.

2. Achieve your stated ROI objectives.

An <u>Effective Maintenance Program</u> that is <u>Efficiently Operated</u> is essential to your success. Bon voyage.

APPENDIX

1. Procedure 1.1.3.12 – Filtration of Lubricating and
    Hydraulic Oils (pages 1 thru 3 and page 1
    Enclosure A)

2. Mechanical and Electrical P/M Procedures and
    Frequencies for Warner & Swasey's 2-SC N/C
    Turning Machine

3. Maintenance Work Order – Procedure 1.1.3.8

4. Blank Weekly Performance Reporting Form

Brown & Sharpe Manufacturing Company	PAGE 1 OF	PROCEDURE	1.1.3.12
AND SUBSIDIARIES	SUBJECT Preventive Maintenance		DATE ISSUED
APPROVED BY:	Filtration of Lubricating and		1-31-72
	Hydraulic Oils in Production		DATE REVISED
	Machines.		
	Applies to: Precision Park		1-12-76

The **purpose** of this procedure is to establish the method and frequency for filtration of lubricating and hydraulic oils on designated critical production machines at Precision Park.

Equipment presently used for this filtration work is the IMP-356 (Imperial Hydraulics) equipped with a Brown & Sharpe pump and having a filtration capacity of 5 gallons per hour. This is a small, portable, electrically operated unit with three-stage filtration.   These filters which are disposable are arranged in series as follows:

> 1st. stage - 100 microns
>
> 2nd. stage - 40 microns
>
> 3rd. stage - 10 microns
>
> 4th. stage -   3 microns

1.   **N/C Machines** will be scheduled for monthly filtration of oil.

2.   Other machines listed in Schedule A will be scheduled for bi-monthly filtration of oil. (Once every two months)

3.   For each **5 gallons** of oil reservoir capacity, the filtration unit will be run for **one hour.**

The objective of this procedure is twofold:

1.   By predetermined oil filtration schedules, keep lubricating and hydraulic oils clean and acid free and thereby extend the oil life to 5 years! (Past practice has been to change oil in critical machines yearly.)

2.   By predetermined oil filtration schedules, keep lubricating and hydraulic oils clean and acid free -- thereby avoiding expensive breakdowns.

Procedure and Frequency

1.   The Maintenance Services Engineer in cooperation with the Department Manager and the Maintenance Foreman will select and list critical production machines in which hydraulic and lubricating oils shall be filtered.

2.   The Maintenance Services Engineer working jointly with Division Manufacturing Engineers and their designated representatives plus technical representatives from the oil vendor will determine the frequency of filtration.

3. The Maintenance Services Engineer will be responsible for scheduling and monitoring oil filtration work. Oil filtration work will be performed by machine repairmen assigned to such work by the Maintenance Foreman. At the beginning of each week, the Maintenance Services Engineer will issue to the Maintenance Foreman maintenance work orders (Form 821-1664) covering filtration of machines to be performed that week. Essential information required to complete the work will be pre-printed or pre-typed on the maintenance work order.

4. The Maintenance Foreman is responsible for maintenance of the IMP-356 oil filtration unit and maintaining an adequate supply of filter elements.

5. The Maintenance Services Engineer is responsible to establish and maintain an adequate file for completed work orders. Initially such completed work orders will be stored for a period not to exceed two years.

6. The Maintenance Services Engineer in cooperation with the Purchasing Department is responsible to arrange for chemical and physical analysis of between five and ten oil samples quarterly with the oil vendor. After the first year, at least half of the oil samples selected for analysis should be from the same production machines from which samples were taken a year ago. Written results of tests performed by the oil vendor should be reviewed with the Maintenance Manager and Division Managers of Manufacturing Engineering or their designated representatives.

7. Enclosure A is a list of critical machines at Precision Park on which oil filtration is to be performed.

8. Division Manufacturing and/or equipment Engineers are responsible for advising the Maintenance Services Engineer of additions or deletions to this list of machines requiring oil filtration.

9. The following list of lubricants are for general use in Brown & Sharpe Manufacturing Company:

   Oils

Designation	ASLE Lubricant Type	B&S Stock No.	Vendor Description	Viscosity @ 100° Unless Specified Otherwise
B&S #8 Oil		840-48	Texaco	S.A.E. #30
B&S #51 Oil	S-32	840-59	Mobil Velocite #3	32 S.U.S.
B&S #52 Oil	S-60	840-60	Mobil Velocite #6	60 S.U.S.
B&S #53 Oil	S-105	840-61	Sunvis 701	100 S.U.S.
B&S #62 Oil	H-150	840-62	Sunvis 706	150 S.U.S.
B&S #63 Oil	W-150	840-63	Mobil Vacuoline 1405	150 S.U.S.

Oils

Designation	ASLE Lubricant Type	B&S Stock No.	Vendor Description	Viscosity @ 100° Unless Specified Otherwise
B&S #64 Oil	H-215 AW	840-126	Sunvis 747	200 S.U.S.
B&S #65 Oil	H-315	840-64	Sunvis 754	300 S.U.S.
B&S #71 Oil		940-65	Sunvis 7100	1500 S.U.S.
B&S #72 Oil	G-2150	840-66	Sunep 150	2150 S.U.S.
B&S #73 Oil		840-125	Texaco	S.A.E. 10W
B&S #81 Oil	W-315	840-67	Sunvis 500 Waylube #80	300 S.U.S.
B&S #82 Oil	W-1000	840-68	Sunvis #90	1000 S.U.S.
		840-133	Dexron H-36	

Review Responsibility: Manager of Maintenance

Distribution: All R. I. Holders of the Management Guide

Review Date: 6/1/79

NUMERICALLY CONTROLLED MACHINES

Dept. No.	Machine No.	Machine Description	Cap. in Gallons	Filtering Time(Hrs.)	B&S Oil Type	Oil Type	Location
Machine Tool Division							
5334	7-137	J&L Lathe	50	10	#62	Sunvis 706	P-14
5334	7-138	J&L Lathe	50	10	#62	Sunvis 706	Q-14
5334	7-139	J&L TNC Lathe	40	8	#62	Sunvis 706	Q-14
5334	7-518	W&S 2 SC	45	9	--	Dexron H-36	P-19
5334	7-519	W&S 2 SC	45	9	--	Dexron H-36	Q-19
5336	7-522	J&L TNC Lathe	40	8	#62	Sunvis 706	R-13
5335	10-178	Hydrotape	40	8	#62	Sunvis 706	N-22
5335	10-179	Hydrotape	40	8	#62	Sunvis 706	N-22
5335	10-236	VRM	50	10	#62	Sunvis 706	O-23
5335	13-64	5" Bullard	50	10	#64	Sunvis 747	M-25
5335	13-65	Sundstrand OM2	50	10	#62	Sunvis 706	O-23
5335	13-66	G&L	50	10	#62	Sunvis 706	P-26
5335	13-67	B&S Hydromaster	42	8	#53	Sunvis 701	N-23
5335	13-68	B&S Hydromaster	42	8	#53	Sunvis 701	N-23
5332	14-102	Behrens Press	200	40	#64	Sunvis 747	O-30

August 29, 1975

Code:
(7-1021)   5.29
(7-519)   5.30
(7-518)   5.31
(7-520)   5.33
(7-2006)   5.34

Preventive Maintenance Schedule
on Warner & Swasey 2-SC
Numerically Controlled Turret Lathes Model M-5040
with Mark Century 7542 Control
Machine No.'s 7-1021, 7-519, 7-518, 7-520 and 7-2006

The following maintenance intervals are suggested for use under optimum conditions. If environmental conditions warrant, the frequency of maintenance intervals should be increased.

Frequency	Service "Electrical"
Weekly	
Monthly	**Air Filters:** Air filters are located on the machine control cabinet. N/C control cabinet. Longitudinal servo drive and D.C. main drive motors. Inspect and clean the air filters with hot water and a detergent or any type of solvent and then blow them dry with air.
Monthly	**Control Panel:**

Control Panel:

1. Check for dirt, oil or water in control panel area and clean.

2. Check for airtight seal of control panel area.

3. Check tightness of screws on terminal boards and relays.

4. Check all switches and operating buttons including signal lights.

3 Month   D.C. Main Motor
Servo Axis Motors
Index Drive Motors and D.C. Tachometers

1. Clean out any accumulated dirt and dust.

2. Check brushes.

3. Check brush spring tension.

246

Frequency	Service "Electrical"
3 Month (cont'd.)	4. Check commutator.
	5. Check mica between commutator bars.
	6. Check coupling tightness on Tachometer of feed-back devices.
3 Month	**Synchro Transmitters and Receivers:**
	1. Clean out any accumulated dirt and dust.
	2. Check brushes for excessive bouncing or arcing.
	3. Check excitation voltage.
6 Month	The flexible magnaloy coupling between the main drive motor and spindle drive transmission requires no lubrication, however, the rubber insert between the couplings should be inspected. Replace if deteriorated or worn.

Frequency	Service "Electrical"
Weekly	Lamp/lens: (Reeler only) Tape guides, reader photocell assembly, guide rollers. Dust, using a soft cloth, cotton swab, or brush. Clean using sparing amounts of pure isopropyl alcohol.
Monthly	Fixed/movable guide roller bearing: Sleeve type: Lubricate with one drop of an oxidation resistant mineral oil such as Gulf A&E or Toreso T-43 or T-52. Apply at the junction of the shaft and bearing. Caution: do not over-lubricate. Roller Type: No lubrication is required for the life of the components.
3 Month	Photocells in reel arm feedback assemblies: (Reeler only) Remove the cover over the photocell assemblies. Clean the faces of the photocells with a cotton swab dipped in pure isopropyl alcohol. Drive motor/reel motors. No lubrication is required for life of the components.
3 Month	Reel motor brushes: Check the length of them and replace them when their length is 5/16 or less. When the brushes are removed for checking, each brush must be replaced in the same holder from which it was removed.
Monthly	Reel shaft assembly: Check for excessive wear causing reel to be loose on shaft. Replace as necessary.
6 Month	Reader/Reeler: Check alignment and adjustments of the following: 1. Light line width and position. 2. Lamp voltage. 3. Solar cell output check. 4. Sprocket wheel.

Code: 5.29
      5.30
-4-   5.31
      5.33
      5.34

Frequency	Service "Electrical"
6 Month (cont'd.)	5. Complete reading head alignment.
	6. Tape tension arm spring.
	7. Reel brake torque and wear check.
	8. Photocell output and balance adjustment.
	9. Reel motor phasing and direction of rotation check.
	10. Motor zone switching points.
	11. Broken tape lever switch actuation points.

6 Month   PWM-Servo System:

Re-tune the servo according to the procedure given in Chapter V.

6 Month   12 volt power supply.
Check and adjust the output.
Voltage should be within 2% of 12V.

1. Check for dirt or water in control panel area.

2. Check for airtight seal of control panel area.

3. Check tightness of screws on terminal boards and relays. (Use CAUTION)

4. Check all switches and operating buttons including signal lights.

Yearly   Resolver Feed Back Unit

Gear Train: Check, clean and lubricate the gear train with a light application of Mor Film Cling oil (250-300 SUS at 100°F).

Synchro: Inspect and replace if necessary.

Torque Motor: Inspect and replace if necessary.

Yearly   Control Cabinet "Cooling Surfaces":

1. Remove the side panels of the case and wipe the inside surface with a dry cloth to remove any dust or dirt accumulated by fans.

Frequency

Yearly (cont'd.)

Service "Electrical"

2. Reach into the air plenum first from one side of the control and then from the other, wiping the fins on top of the control until they are clean.

3. Clean the internal walls of the square tubes forming both side panels. Swab with a clean, dry cloth to remove accumulated dust and dirt. Since these surfaces are the least likely to cause trouble, this step is the final one to be taken.

V. S. Blair

Code:
(7-1021)   8.29
(7-519)   8.30
(7-518)   8.31
(7-520)   8.33
(7-2006   8.34

Preventive Maintenance Schedule
on Warner & Swasey 2-SC
Numerically Controlled Turret Lathes Model M-5040
Machine No.'s 7-1021, 7-519, 7-518, 7-520 and 7-2006

The following maintenance intervals are suggested for use under optimum conditions. If environmental conditions warrant, the frequency of maintenance intervals should be increased.

Frequency	Service "Mechanical"
Weekly	Check main hydraulic reservoir.
	Check slide way lubrication reservoir for level.
	Check cleanliness of ways and ball screws.
	Check oil level in air panel.
	Grease air cylinder packing gland.
	Grease chuck.
	Drain condensation from air panel.
	Check for leaks at hoses, tubings and fittings.
	Clean machine thoroughly and wipe down.
	Note: Do not use air for cleaning purposes. The pressure will drive dirt and chips into bearing surfaces.
Monthly	Check operation of mechanical interlocks.
	Check for grease leaks in motor and oil leaks from housing.
	Chip tote: Check oil level in reducer.
	Air control panel. Remove the filter, clean out the accumulated dirt and rust particles.

Frequency	Service "Mechanical"
3 Month - Millwright	Chip Tote: Conveyor Inspect track and roller chain for sign of wear. Make sure rollers turn freely, lubricate if necessary. Chip Tote: Clutch Inspect clutch and adjust if necessary. Chip Tote: Drive Unit Inspect-make sure the "V" belts and chain are aligned and slightly loose. Chip Tote: Lubrication Lubricate belt chain. Grease pillow blocks. Change oil is speed reducer. Oil drive chain.
6 Month	Main oil reservoir: Change oil and filters. Clean and wash all way wipers and check wear condition. Replace if worn. Main drive coupling. The flexible magnaloy coupling between the main drive and spindle drive transmission requires no lubrication, however, the rubber insert between the coupling should be inspected. Replace if deteriorated or worn.
Yearly	Replace way wipers as needed with special attention to wipers adjacent to metal cutting area. Turret Index Drive. Check level of grease in index drive transmission and repack with fresh grease. Gib and Caps: There are three tapered gibs between the bedways and the cross slide. One under each way is held in place by a bottom cap and a turret slide gib between the cross slide and the rear side of the right way. The gib setting should be checked for correct adjustment as part of a regularly scheduled preventive maintenance.

Frequency   Service "Mechanical"

Yearly (cont'd.)   Lubricate bearings in D.C. Main
Drive Motor.

Check level of grease in index drive
transmission and repack with fresh
grease per w/s lubrication specification.

V. S. Blair

Brown & Sharpe Manufacturing Company
AND SUBSIDIARIES

APPROVED BY:

PAGE 1 OF 2   PROCEDURE   1.1.3.8

SUBJECT
DATE ISSUED
4-1-68

MAINTENANCE WORK ORDER

DATE REVISED

Applies to: Precision Park Only   2-15-78

The <u>purpose</u> of this procedure is to provide an effective and economical means for planning, scheduling, and dispatching maintenance work. Maintenance Work Order Form No. 821-1664 will be used with this procedure.

This procedure will apply where the cost of maintenance work required does not exceed fifty (50) hours for labor or $500 for material. When either of these limits will be exceeded, the Work Order Form 457 will be initiated and approved per present practice.

Any foreman or department supervisor requiring maintenance work should initiate Maintenance Work Order Form No. 821-1664. The information to be entered by the originator is shown on the following sample:

B & S MAINTENANCE WORK ORDER FORM	TIME NUMBER OR EXPENSE ACCOUNT #: 5334-317-316-12	ORDER WRITTEN	
		TIME	DATE
		10:30 a.m.	4-1-68
	DEPT. / SECTION	LOCATION	DESCRIPTION OF MAINTENANCE
	EXPECTED COMPLETION DATE: 4-3-68		REPAIR DEFECTIVE CLUTCH ON MAIN DRIVE.
	ACTUAL COMPLETION DATE		
	TIME	DATE	
			SIGNATURE: N. E. SMITH

FORM 821-1664

The originator will retain the original (white paper copy) and deliver the remaining two copies to the Maintenance Department.

Maintenance Work Orders involving installation of new equipment, relocation of existing equipment, and/or removal of old equipment will originate in the Plant Engineering Office. These Maintenance Work Orders will result from receipt of Machinery and Equipment Installation or Transfer Slips (Form 278) from the appropriate Division Director of Manufacturing Engineering or his designated representative. It is requested that the following lead time schedule be used in the issuance of these Transfer Slips:

1. Installation of standard equipment not involving replacement of existing equipment.   fifteen (15) days

2. Installation of standard equipment involving replacement of existing equipment.

thirty (30) days

3. Installation of equipment requiring special electrical or electronic controls.

Consult with Plant Engineering allowing lead time to purchase long lead electrical components.

4. Installation of equipment requiring a foundation.

ninety (90) days

Upon receipt of the Maintenance Work Order, the Maintenance Clerk will stamp the date and time received on the second and third copies. The appropriate Maintenance Section Foreman will check the expected completion date entered by the originator. Where the expected completion date cannot be met, the appropriate Maintenance Section Foreman will contact the originator to reach U & A on a different completion date. In a few cases, U & A may have to be reached at a higher management level.

The estimated time prepared under the direction of the Maintenance Industrial Engineer will be used for maintenance work load determination by trade.

The Maintenance Clerk will advise the originator of the order that the job has been completed, and will mail the second copy to the originator showing the total actual time spent.

For maintenance work estimated to require less than 0.2 hours, the Foreman or Department Supervisors should call the Maintenance Department and make necessary arrangements with the proper Maintenance Section Foreman. For these few jobs, no Form No. 821-1664 will be prepared.

Review Responsibility: Manager of Maintenance

Distribution: Corporate, Machine Tool, PTG&P

Review Date: 6/1/79

INDEX